THE TITANS OF TECH

THE TITANS OF TECH

EDISON TO GATES

For general information on our other products and services or to obtain technical support, please contact our Customer Care Department within the United States at (866) 744-2665, or outside the United States at (510) 253-0500.

Lightning Guides publishes its books in a variety of electronic and print formats. Some content that appears in print may not be available in electronic books, and vice versa.

ISBN Print 978-1-942411-35-2
eBook 978-1-942411-36-9

Front Cover Photo: © DARPA/Boston Dynamics
Back Cover Photo: Javier Pardina/Stocksy
Interior Photos: ACALU Studio/Stocksy, p.1; © Hulton-Deutsch Collection/CORBIS, p.2; Tony Avelar/Bloomberg/Getty Images, p.5; Meyer Pfundt/Getty Images, p.6; PAUL SAKUMA/ ASSOCIATED PRESS, p.7; © scanrail/iStock, p.9; © FotoSpeedy/iStock, p.9; Mega Pixel/ Shutterstock, p.9; asharkyu/Shutterstock, p.9; fad82/Shutterstock, p.9; tovovan/Shutterstock, p.9; Skocko/Shutterstock, p.9; Science & Society Picture Library/Getty Images, p.13; Steve Cukrov/ Shutterstock, p.14; © SuperStock/CORBIS, p.17; © GraphicaArtis/CORBIS, p.22; © CORBIS, p.23; Encyclopaedia Britannica/UIG/Getty Images, p.24; Africa Studio/Shutterstock, p.25; © USDA Photo/Alamy, p.27; Museum of Science and Industry, Chicago/Getty Images, p.28; © Corbis, p.32; © Corbis, p.34; Library of Congress/Getty Images, p.37; © H. Armstrong Roberts/CORBIS, p.38; Lucidio Studio Inc, p.41; PhilipYb Studio/ Shutterstock, p.42; © D&P Valenti/ClassicStock/ CORBIS, p.45; Kathryn Greenhill [CC BY-SA 2.0]/ Wikimedia Commons, p.46; oldmonk/Shutterstock, p.49; Ted Streshinsky/The LIFE Images Collection/ Getty Images, p.50; Courtesy of Intel, p.53; Brad Trent, p.55; JOKER/Martin Magunia/ullstein bild/ Getty Images, p.56; Twin Design/Shutterstock, p.57; © Matthew McVay/CORBIS, p.59; ©James Leynse/CORBIS, p.62; drserg/Shutterstock p.63; © giuseppe masci/Alamy, p.65; CHARLES PLATIAU/ Reuters/CORBIS, p.66; DWD-Media/Alamy, p.67; Tim Boyle/Bloomberg via Getty Images, p.68; netopaek/Shutterstock, p.68; MaleWitch/ Shutterstock, p.69; Cate Frost/Shutterstock, p.69; © Anthony Brown/Shutterstock, p.69; © Kim Kulish/CORBIS, p.72,73; © Ian Dagnall/ Alamy, p.74; Nikola Spasenoski/Shutterstock, p.77; Karramba Production/Shutterstock, p.79; Christoph Dernbach/picture-alliance/dpa/AP Images, p.81; Norm Betts/Bloomberg News/Getty Images, p.83; © Aflo Co. Ltd./Alamy, p.84; Feng Yu/Shutterstock, p.85; © Martin Klimek/ZUMA Press/CORBIS, p.87; Andrew Harrer/Bloomberg/ Getty Images, p.90; Tony Cordoza/Getty Images, p.91; © Izel Photography - Snapshot/Alamy, p.95; Alejandro Moreno de Carlos/Stocksy, p.96,97; MAHATHIR MOHD YASIN/Shutterstock, p.99; Rawpixel/Shutterstock, p.100; Christopher Meade/ Shutterstock, p.101; Alexander Raths/Shutterstock, p.104; © AMELIE-BENOIST/BSIP/BSIP/CORBIS, p.106; Courtesy of DARPA, p.107; Andrew Zarivny/ Shutterstock.com, p.108

CONTENTS

FROM THE EDITOR

**“Hell, there are no rules here—
we’re just trying to accomplish something.”**

—THOMAS EDISON

So spoke Thomas Edison, the “Wizard of Menlo Park,” expressing a sentiment echoed in a thousand incarnations by technical giants throughout history. Henry Ford, Bill Gates, Eli Whitney, Johannes Gutenberg, Leonardo da Vinci. These were the rule-breakers. The visionaries. The ones who saw further, dreamed bigger. The ones who dared. They did it for fame, for profit, for the love of knowledge and for the thrill of discovery. Humanitarians, tyrants, geniuses, and *enfants terribles*—it is their vision and sweat, avarice and sacrifice, that has defined our future. As J. Robert Oppenheimer, father of the atomic bomb, once said, “Both the man of science and the man of action live always at the edge of mystery, surrounded by it.” We can only look on, in awe or terror, marveling at what these titans of technology have wrought.

Google
B44
Lobby

INTRODUCTION

e are all end users of technology. The book you're reading right now—whether it's a paperback or an e-book—is a form of technology. So are contact lenses, rubber wheels, and radio waves. In our daily lives, we interact with more technology than we can ever fully appreciate. Many people are okay with that—they're happy to enjoy their iPhones and zippers and shoelaces without a second thought.

But the end user is only the end of the story. For every iPhone—for every component of every iPhone—there is a person who put the pieces together, and a person who dreamed up the pieces in the first place. Some of these titans of technology are visionaries, like Elon Musk, who cofounded PayPal and Tesla Motors. Others are little-known people who simply recognized a solution to a common problem, like Josephine Cochran, who invented the dishwasher in 1886. Every technology has an origin story. Together these stories paint a vivid picture of human resourcefulness and ingenuity.

You can use a light bulb to brighten a room. Or you can understand the light bulb itself—who first imagined it, what problems it solved, how it was manufactured and adopted by an entire society—and illuminate the world around you.

FIRST, A FEW FACTS

1933

EDWIN HOWARD ARMSTRONG
TUNES US INTO
FM RADIO

1954

invention of the first
SOLAR CELL

At NASA in 1980, Valerie Thomas patents

THE ILLUSION TRANSMITTER
AN EARLY FORM OF
3D TECHNOLOGY

JAMES FERGASON
INVENTS LIQUID CRYSTAL DISPLAY (LCD)

1971

How many patents did Thomas Edison have?

The American inventor and businessman earned a remarkable 1,093 during his lifetime. A testament to his prolific achievements, Edison held the record for the most patents earned by a single person until 2003, when Japanese inventor Shunpei Yamazaki took the record. The phonograph, the incandescent light bulb, an early motion picture camera, the carbon microphone, and the rechargeable nickel-vinc battery are among Edison's many groundbreaking inventions.

Did Alexander Graham Bell found Bell Labs?

No. Alexander Graham Bell founded the Volta Laboratory in Georgetown, Washington, DC in the 1880s, which later became the Bell Laboratory. Bell Labs was established in 1925 by Western Electric and AT&T. Both labs share the namesake of Alexander Graham Bell.

What was Al Gore's role in the invention of the Internet?

Al Gore sponsored the 1991 High-Performance Computing and Communications Act, also known as the "Gore Bill," which allocated a hefty amount of money to high-performance computing. This government push was a crucial contribution in stimulating the development of the Internet. Gore also coined the term "Information Superhighway," used to refer to the telecommunications infrastructure that would become the Internet.

What was the only book on Steve Jobs' iPad when he died?

According to Walter Isaacson's Steve Jobs biography, the only book that Jobs downloaded on his iPad 2 was *Autobiography of a Yogi* by Paramahansa Yogananda, a work that held particular importance to Jobs throughout his life. Isaacson notes that Jobs first read the Yogananda's guide to meditation and spirituality in high school. He revisited it while he was in India, and went on to re-read it every year until his death.

Who are the most powerful women in technology today?

While this question leaves plenty of room for debate, both *Forbes* and *Fortune* publish a list of the world's most powerful women every year, and 2014 lists from both magazines include many female leaders in technology. Women in tech who appear in the top 25 of both lists include Sheryl Sandberg, COO of Facebook and the first female member of Facebook's board of directors; Virginia "Ginni" Rometty, Chairman, President and CEO of IBM; Safra Catz, Co-President and CFO of Oracle; Susan Wojcicki, CEO of YouTube; Marissa Mayer, CEO of Yahoo!; Meg Whitman, president and CEO of Hewlett-Packard; and Ursula Burns, Chair and CEO of Xerox and the first African-American woman to lead a Fortune 500 company.

TWENTIETH CENTURY AMERICAN INNOVATION

FROM THE WRIGHT BROTHERS TO NASA ENGINEERS

From the Wright Brothers taking the first flight in 1903 to the now commonplace luxury of flying across the United States in six hours, the once-impossible dream of human flight has certainly come a long way.

Today, the average American books her travel plans online, uses e-tickets to check in for flights, and for a growing number, displays a digital boarding pass on her mobile phone before getting on a plane. None of this would have been possible without the efforts of pioneering Americans whose discoveries brought technological advancement in our everyday lives.

In 1914, World War I brought airplane technology to the fore-front of US consciousness. The National Advisory Committee for Aeronautics (NACA) was formed in 1915 to institutionalize

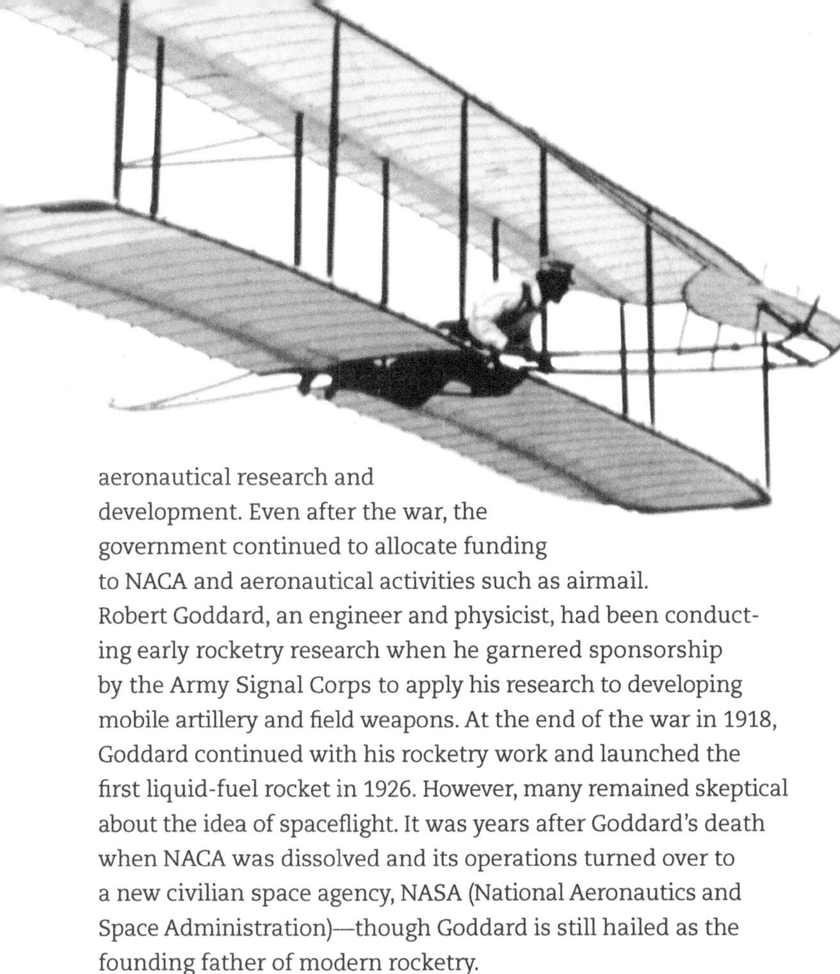

aeronautical research and development. Even after the war, the government continued to allocate funding to NACA and aeronautical activities such as airmail. Robert Goddard, an engineer and physicist, had been conducting early rocketry research when he garnered sponsorship by the Army Signal Corps to apply his research to developing mobile artillery and field weapons. At the end of the war in 1918, Goddard continued with his rocketry work and launched the first liquid-fuel rocket in 1926. However, many remained skeptical about the idea of spaceflight. It was years after Goddard's death when NACA was dissolved and its operations turned over to a new civilian space agency, NASA (National Aeronautics and Space Administration)—though Goddard is still hailed as the founding father of modern rocketry.

Space exploration has always been NASA's core mission, but its research has also made instrumental contributions to everyday American life. Vanguard 2, the first weather satellite, was launched by NASA in 1959. Its successor—TIROS-1—emerged

---

Above: Airplane co-inventor Wilbur Wright glides down the north slope of Big Kill Devil Hill near Kitty Hawk, North Carolina, in a double-ruddered glider.

a year later and was hailed as the first effective weather satellite. Meteorologists rely on these satellites to make the most accurate and up-to-date weather forecasts. They use the technology to monitor impending natural disasters, too, allowing people to make preparations and head to safety. Another NASA contribution has been the satellite navigation system, of which an early prototype, the TRANSIT-1B, was successfully launched in 1960. This system was a precursor to the global positioning system (GPS) that commuters and travelers rely upon today. Finally, NASA is also responsible for launching Telstar, the first communications satellite designed to transmit telecommunications and high-speed computer data and radio and television broadcasts. The COMSAT (short for communications satellite) was invented by aerospace engineer John Robinson Pierce in 1962, and it makes everything from international text messaging to watching our favorite HBO shows possible.

Communications technology hit its stride in the 1960s and '70s. By 1969, scientist George Sweigert had patented the cordless telephone. This new technology could eliminate many missed calls since it allowed people to answer a ringing

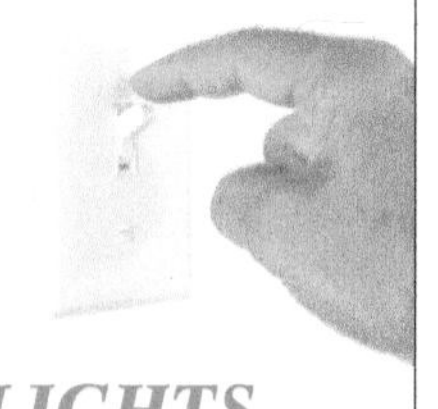

LIGHTS ON!

In 1917, the toggle light switch was invented by William J. Newton. Today, we still use this switch to operate electric lights and electrical outlets. Remote control switches and automatic dimmers have also been added to the mix.

phone more quickly. Motorola engineer Martin Cooper pushed this development many steps further when he lead the first team of scientists to bring personal mobile phones to market. After patenting his "radio telephone system" in 1973, Cooper used his prototype, the DynaTAC, to call his competitor at Bell Labs, Joel Engel, in front of a captive audience in Manhattan, New York. It took Cooper 10 years before his team gained FCC approval in 1983 and released their phone to the market in 1984. Nowadays, still hold audiences captive, but for very different reasons than in 1973. Today's cell phones offer so many features, some people experience separation anxiety when their phones are not within reach.

When Thomas Edison invented the rechargeable nickel-zinc battery in 1901, no one could have predicted the life-changing impact it would have. Mobile phones, digital cameras, Bluetooth speakers—even cars by American companies such as Tesla Motors and Chevrolet are rechargeable these days. Prior to batteries, people were leashed to their devices via two other crucial 20th century inventions, the AC power plug and outlet. Invented by Harvey Hubbell in 1904, the power plug was conceived when Hubbell saw a janitor detach and reattach many power cords as he cleaned around electrically operated games at an arcade. Hubbell's plug and outlet, combined with Edison's rechargeable battery, have made such inconveniences a thing of the past.

Many inventions are birthed from a need for convenience and lead to entrepreneurial success—a narrative that echoes throughout American history. Marion Donovan was dissatisfied with the mess of her daughter's cloth diapers, so she created a waterproof diaper using her shower curtain. She called it the "Boater" and sold it at Saks Fifth Avenue before selling her patent to Keko Corporation for $1 million—an astronomical price tag in 1951. From then on, she continued to collect patents for convenience-minded products such as a facial tissue box, a hanging closet organizer, and a soap dish that drains into the sink. Donovan's legacy is a testament to American entrepreneurism.

There are plenty of modern inventions we take for granted every day. Take Otto Rohwedder's bread slicing machine for instance—a simple idea that is nearly synonymous with innovation. (Consider how often something is heralded as "the best thing since sliced bread!") Today a loaf of sliced bread is a staple on most grocery lists, as are rolls of another mundane invention: paper towels. First marketed by Scott Paper Company in 1907, paper towels became a wildly popular item for kitchens and bathrooms. In 1948 George Clemens invented the hand dryer, which helped cut down on waste, and contributed to environmental sustainability.

In 1941 at General Electric, George Inman developed the first practical fluorescent lamp—a light source that used 70 percent less energy than incandescent lamps. Today, fluorescent lighting remains crucial to energy conservation efforts. However, it doesn't hold a monopoly on energy-efficient lighting. Nick Holonyak, Jr.'s 1962 invention—the light-emitting diode, more commonly known as LED—shares the market with fluorescents, offering a longer lifetime, though at a higher cost.

The 20th century was a time of prolific change across the American landscape. Much of the technology developed defines the way we live today. From products of convenience communication technology that influences us at on an unprecedented level, innovation has catapulted our culture and society into a future once only dreamed of in science-fiction novels.

AMERICA GETS A COOLER OPTION

WILLIS CARRIER'S AIR APPARATUS

Imagine a time when excessive heat and humidity would interfere with worker productivity and manufacturing processes at a national level, when no one could escape suffocating heat. This was the norm for the American workforce before Willis Carrier invented modern air conditioning. It was a foggy day in 1902 when Carrier stood on a train platform in Pittsburgh and realized he could run air through cold water to create controllable humidity—the basis of his design for his "Apparatus for Treating Air." Carrier was dubbed "a Johnny Icicle planting the seeds of climate control all across America," by *American Heritage* magazine.

Carrier's work led to his takeover of the engineering department at Buffalo Forge, an industrial machine manufacturer. By 1907, Buffalo Forge had embraced the new industry that Carrier was pioneering and launched a subsidiary: Carrier Air Conditioning of America. Textile plants became the first to adopt the technology. Air conditioning quickly increased worker productivity by creating a more comfortable work environment and making textiles easier to process. Pharmaceutical plants also embraced the technology. Carrier then procured a contract with the Celluloid Company, which made film for the motion picture industry. It wasn't long before other industries used air conditioning to boost the production of goods such as rubber, rayon, razor blades, and flour.

At the start of World War I, factories and corporations harbored fear over the eventual outcome of the war. Buffalo Forge decided to shutter Carrier Air Conditioning of America. That didn't stop Willis Carrier from continuing with his vision. With Irvine Lyle, he started the Carrier Engineering Corporation in 1915. They continued to lend their engineering expertise to the Allied war effort.

When Carrier introduced the economical centrifugal chiller in 1922, large-scale air conditioning began to appear in more public and commercial settings: office buildings, department stores, and movie theaters. Sid Grauman's Metropolitan Theatre in Los Angeles was the first to install bypass circulation and

Willis Carrier discovered the "law of constant dew-point depression," stating, "constant dew-point depression provided practically constant relative humidity." The dew point is the temperature at which the water vapor in air condenses into water and simultaneously evaporates.

down-draft distribution systems, which allowed moviegoers to enjoy a gentle flow of air from ceiling vents. By the time Carrier began installing air conditioning systems in the Palace Theatre in Dallas and The Texan in Houston, centrifugal chillers had become fully integrated. A complete centrifugal chiller system in the Rivoli Theatre in New York was Carrier's pièce de résistance of this period, though the practical applications for air conditioning continued to grow. In 1925, Carrier was commissioned by the US Navy to air condition the engine room of the destroyer USS *Wyoming*. That same year, the brand-new Madison Square Garden in New York requested Carrier's services, too. The venue's management had plans to produce an ice surface for hockey, and wanted Carrier to help them achieve it.

At the dawn of the Great Depression, Carrier Engineering Corporation merged with Brunswick-Kroeschell Co. and York Heating & Ventilating Corp. and became Carrier Corporation. During this time Carrier worked on a steam-ejector refrigerating system for the railroad industry. He had discovered that steam could be used to cool water and thus created an efficient way of cooling railcars. This garnered greater public acceptance of Carrier's work. After additional projects with military ships and luxury cruise liners, Carrier ventured into the health care industry. A New York hospital utilized a Carrier cold diffuser to cool its cryotherapy room where cancer patients could receive cold temperature treatment to repair damaged tissues.

It wasn't long before Carrier saw the need for standalone air-conditioning units. In 1931, the Atmospheric Cabinet—a cabinet with a fan and cooling coil and a remote refrigerating machine—was introduced as a personal or small-room cooling system. The company continued to develop its concept of "manufactured weather" in the home, developing another product called the Carrier Room Weathermaker. Simultaneously the company continued its large-scale work, serving Rockefeller Center in New York and the Pentagon in Washington, D.C. The Conduit Weathermaster System, which allowed for moisture-controlled air to travel through narrow ducts quickly and used individually controlled units to heat or cool the air, was developed for large, multi-unit buildings.

What started as an idea conceived on a foggy train platform turned homes, cars, and even airplanes into comfort zones.

Carrier again contributed to the US war effort in World War II. The military utilized the company's refrigerating systems for food preservation and called upon its general engineering prowess to produce classified equipment. In a unifying, patriotic act, many department stores across the country that had benefitted from Carrier's invention—big names like Macy's, Tiffany & Company, Lord & Taylor, Gimbels, Sears, Marshall Field's, and J. L. Hudson—gave up their air-conditioning systems and donated them to companies such as B. F. Goodrich, who were ramping up production on materials for the war.

Willis Carrier passed away in 1950, but his legacy lives on. In 1953 Carrier Corporation sold more than 1 million room air-conditioning units in the US residential market. By 1962, heating and cooling were becoming a public utility, starting in Hartford, Connecticut. The Hartford Gas Company provided air conditioning and heat to downtown residents via underground pipelines. The spread of Carrier's air conditioning technology fueled population growth and spurred a migration of Americans to the "Sun Belt" in the southeastern and southwestern United States.

Large commercial spaces remained an important part of Carrier's interests. Though Carrier had installed systems in skyscrapers in the past, nothing compared to the company's work with the Twin Towers of the World Trade Center in New York in 1969—that is, until Carrier was contracted to air condition the 110 floors of the Sears Tower that were under construction in Chicago in 1972. In the early 1980s, Carrier served Disney World's EPCOT Center and City Place in Hartford, Connecticut—symbolizing the company's leap into the future. Dubbed the world's first intelligent building, City Place used advanced computers to operate building services. To perform well, the computers needed air conditioning to maintain a temperate environment.

What started as an idea conceived on a foggy train platform turned homes, cars, and even airplanes into comfort zones in themselves; as well as malls, department stores, and smaller shops, which vitalized the economy. The hot summer months became the most profitable season for the American film industry once air conditioning gave people in movie theaters a retreat from the heat. This generated the booming business of summer movie blockbusters. Willis Carrier's invention has made a lasting impact across a myriad of aspects of modern life in the United States and beyond.

MORSE CODE

SAMUEL MORSE AND THE ELECTRIC TELEGRAPH

Before Samuel F. B. Morse developed his popularized version of the telegraph in 1837, intellectuals in Germany, England, and Russia had already developed various types of electrical telegraphs. There were the electrochemical telegraph by Samuel Thomas von Soemmerring, the electrostatic telegraph by Francis Ronalds, the electromagnetic telegraph by Baron Schilling, and the electromechanical telegraph by Carl Friedrich Gauss and Wilhelm Weber. The first commercial version of the technology was the Cooke and Wheatstone telegraph, developed by the British scientists William Cooke and Charles in May 1937. Theirs would soon be eclipsed by Morse's cheaper telegraph, but not before Morse spent some time pursuing his other passions.

Morse was a well-educated socialite, having studied religious philosophy, mathematics, and science at Yale College. In 1811, he followed his dreams and traveled to London to study art and portraiture at the Royal Academy. When Morse returned to the United States after the War of 1812, he became a prolific painter, exploring subjects of American culture, history, and religion, as well as painting portraiture. By 1826, he had cofounded the National Academy of Design in New York City and was serving as its first president.

According to popular myth, the invention of Morse's telegraph was born out of tragedy. In 1825 while Morse was in Washington,

DC, he received a letter from his father, via horse messenger, informing him that his beloved wife, Lucretia Pickering Walker, had fallen gravely ill. By the time Morse reached New Haven to be at his wife's side, she had already been laid to rest. Many believe that Morse was heartbroken that he received the news of his wife's illness so late, and it was this grief that led him to invent the telegraph.

In reality, it's more likely Morse thought of the telegraph after he met physician and scientist Charles Thomas Jackson in 1832. Jackson's knowledge of electromagnetism sparked Morse's interest and led him to design a single-circuit telegraph comprising wooden clock wheels, wooden drums, levers, cranks, paper rolled on cylinders, a triangular wooden pendulum, an electromagnet, a battery, and copper wires. A peculiar device, it was essentially a key, a battery, and a wire connected to an electromagnetic receiver. By pushing the key down to complete the electric circuit of the battery, an electrical current would travel across the wire to the receiver. To transmit actual data through this system, Morse and his assistant Alfred Vail developed Morse code, which assigned a series of dots (short pulses) and dashes (long pulses) to each letter and number. These dots and dashes would be transmitted through the system and then translated by an operator on the receiving end. It was a

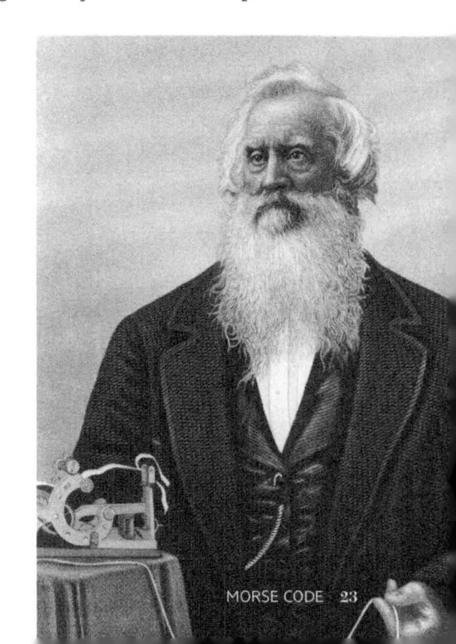

> AND THE EXCLAMATIONS, 'EXTRAORDINAIRE!' 'TRÈS BIEN!' 'TRÈS ADMIRABLE!' I HEARD ON ALL SIDES. —SAMUEL F. B. MORSE

simple yet genius concept. Morse sent the first telegram in the United States in 1838. In 1844 he held a public demonstration of his technology from the Capitol in Washington, DC, sending a message to the old Mount Clare Depot in Baltimore: "What hath God wrought?" By 1861, the system was widely used and facilitated quick bicoastal communication in the United States.

Morse's more economical system led to a phase-out of the Cooke and Wheatstone telegraph system. Telegraph operators developed an ability to understand Morse code aurally, which improved the efficiency of message relay. An inventor named Royal Earl House developed a letter-printing telegraph in 1846, reducing the chances for human error. The Western Union Telegraph Company was the first to create a unified system of telegraph stations in the 1850s. By 1861 the company had gone nationwide. Journalists embraced the technology, changing the way news was spread. By the 1890s, telegraphy was wireless and Morse code was being utilized in early radio communication. The US military began experimenting with the technology after the turn of the century. However, it wasn't until the 1930s that knowledge of Morse code would become a standard requirement for military pilots and civilians. It was a crucial tool during World

War II, allowing for fast and secure communications.

There has always been controversy over Samuel F. B. Morse's achievements with the electric telegraph. Without a doubt, he was inspired by the brilliant people around him, but rumors about Morse exploiting their ideas dogged him. He was known to have been helped by a New York University professor of chemistry, Leonard Gale, and Charles Jackson later claimed that Morse borrowed from his own theories of electromagnetism. Alfred Vail played a key role in Morse's work, although he seemed to relinquish much of the credit to Morse. Morse was extremely protective of his claims regarding the invention of the electric telegraph. He actively engaged in the race to create the gold standard of the technology and was outspoken in refuting others' criticisms and appropriations.

Telegraphy was a harbinger of the social networking phenomenon to come in the 21st century; Morse operators engaged in their own form of social media, using the system to communicate personal life events to one another. In a way, the telegraph paved the way for new media and modern communication tools.

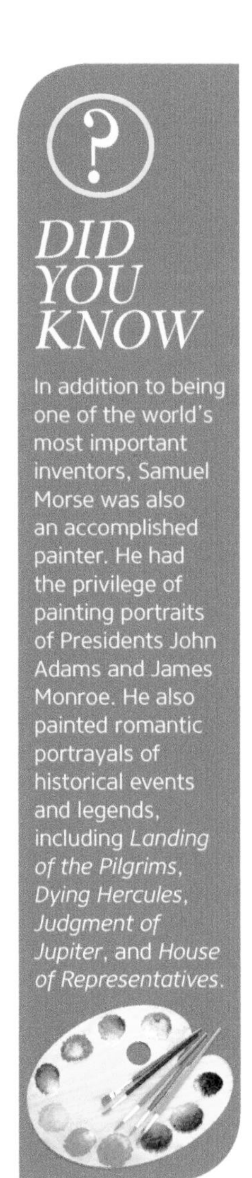

DID YOU KNOW

In addition to being one of the world's most important inventors, Samuel Morse was also an accomplished painter. He had the privilege of painting portraits of Presidents John Adams and James Monroe. He also painted romantic portrayals of historical events and legends, including *Landing of the Pilgrims*, *Dying Hercules*, *Judgment of Jupiter*, and *House of Representatives*.

CLOUD OF CONTROVERSY

ELI WHITNEY AND ALEXANDER GRAHAM BELL

IT was a time when patent laws were still in their infancy and the intellectual property of inventors was vulnerable. Though innovation drove economic growth for beneficiaries of inventions like the cotton gin and telephone, it did little to protect the interests of their inventors.

The institution of slavery in America was in decline by 1790. Tobacco farming had depleted much of the soil in the south, and demand for that crop was on a downswing; rice, indigo, and corn were no longer as profitable, either. The American consciousness was heading toward an end to slavery, with some plantation owners having released slaves already or pledging manumission upon the owners' deaths.

Eli Whitney was a man gifted with an instinct for mechanism. In his youth he worked as a blacksmith and developed his own nail manufacturing machine before attending Yale College in 1789, against his working-class parents' wishes. After graduating, he headed south and met the widow of General Nathanael

Right: A cotton gin in operation, 1940.

Greene, Catherine Littlefield Greene. She convinced him to come to her plantation in Savannah to work with her fiancé, and Whitney's future partner, Phineas Miller. His arrival would change the course of history in the Antebellum South and throughout the nation.

At the Mulberry Grove plantation, Greene and her neighbors discussed the discouraging state of their crops. Green-seed cotton was the only variety that could grow in their region, but the process of extracting the lint from the seed was tedious and inefficient; it took 10 hours to work on three pounds of seeds. Greene and Miller asked Whitney if he could devise a solution to their productivity problem. Whitney observed how laborers extracted cotton fibers by holding a seed in one hand and teasing out the lint with the other. He mimicked the process by putting the seeds onto a wire sieve and using a rotating drum with hook-shaped wires to graze the sieve and catch the lint. A rotating brush collected the lint off the hooks. With that, the cotton gin was born.

THE GIN

A cotton gin rapidly separates cotton fibers from cottonseeds allowing for greater efficiency than manual separation. Handheld roller gins had been in India and other countries since at least AD 500. However, Whitney's combination of a wire screen and small wire hooks to pull the cotton revolutionized the cotton industry.

{ **Patent caveat:** descriptions, designs, and drawings that serve as official notice of intention to file a patent application }

All of a sudden, a single slave could use the cotton gin to do in one hour what used to take several slaves a full day. Word traveled fast about Whitney's invention, and plantation owners all over the region quickly began planting fields and fields of green-seed cotton. The demand for the cotton gin erupted. Cotton covered the landscape, and Whitney could not keep up with production demands, so plantation owners took matters into their own hands, going as far as breaking into Whitney's workshop and copying his model. He was granted a patent in 1794, but it would not be validated until 1807.

Whitney had intended to relinquish individual sales of the machines and ask for a percentage of the profits earned from cotton sales. This was meant to be a reasonable arrangement—he would make the cotton gin public property. But with the massive proliferation of the crop, millions of dollars were at stake, and people began pirating the gin. By 1797, Whitney and Miller were out of business. Six years later, they were being sued by the Southern states. In the end, Whitney earned about $90,000 off of his invention while plantation owners turned a collective profit of almost $10 million in that year alone.

Southerners were excited by sales of their "King Cotton!" Their crops made up 75 percent of all cotton produced in the world. But with increased mass production came an increased need for more labor. While the cotton gin was invented to reduce labor, its success actually intensified the need for it. More slaves were needed to grow and pick the crop. The number of slaves went from 700,000 in 1790 to 3.2 million by 1850. Some believe that the advent of the cotton gin—and its impact on slavery—was one of the principal factors leading to the American Civil War in 1861.

By the end of the Civil War, new technology was improving communication, and the researchers fueling this change were in fierce competition with each other. One of the most disputed—and complicated—examples of this was the invention of the telephone. Alexander Graham Bell has been widely regarded as the inventor. Having lived a life profoundly touched by deaf people—including his own mother, wife, and students at the Boston School for Deaf Mutes, where he taught—Bell was fascinated by the principles of acoustics. Before immigrating to Canada and the United States from the United Kingdom, he witnessed a demonstration of an early telegraph developed by Charles Wheatstone in London.

By 1874, Bell was working on a harmonic telegraph, which sent acoustic messages at varying pitches through a single wire. Bell soon hit a wall because he lacked the expertise needed to build a working model of his device and conduct the necessary experiments. When he met an electrical designer and mechanic named Thomas A. Watson, he found a partner who could help him realize his vision. Bell theorized that undulating electrical currents could be converted into sound waves. With Watson's help, they were able to achieve bidirectional transmission of speech through vibrating steel reeds.

Bell obtained the patent for the telephone in March 1876. He debuted the technology at the Centennial Exhibition in Philadelphia on June 25, 1876, reciting Hamlet's soliloquy over the phone line to the visiting Emperor of Brazil. Bell and his partners offered the patent to Western Union for only $100,000, but the telegram company had hired Thomas Edison and Elisha Gray to develop similar technology. The Bell Company was then formed in 1877. When Western Union tried to revisit the offer two years later, Bell was no longer interested in selling.

Elisha Gray had filed a patent caveat for an acoustic telegraph similar to Bell's invention on February 14, 1876. Purportedly, Bell's lawyers had filed a similar patent hours after Gray but managed to procure priority for their claim over Gray's. Controversy ensued over whether Bell or his lawyers unethically obtained access to Gray's patent caveat and plagiarized a portion of it. After years of litigation battles and extensive documented evidence of Bell's work, the courts ruled in favor of Bell, and he remained legally recognized as the inventor of the telephone.

However, there was another contender who had been overlooked throughout all of this. An Italian immigrant named Antonio Meucci had created a version of the telephone—called the *telettrofono*—that he had been using in his New York home since 1856. It was used to communicate from his basement laboratory to his sickly wife in the bedroom. Reportedly, a lack of funding had prevented Meucci from upgrading the patent caveat he filed in 1871 to a patent. By the time he died in 1889, he had come close to having his case heard by the Supreme Court. In 2002 the House of Representatives granted recognition of Meucci's work in the development of the telephone.

RIDING THE WAVE

REGINALD FESSENDEN AND RADIO TECHNOLOGIES

Imagine a rock dropping into a pool of water. Ripples circle out from the point where the rock plops through the water's surface. Think of that point of impact as an antenna, and the ripples as electromagnetic waves carrying sound. Those waves must continue to radiate steadily until they reach another receiving antenna in order to transmit the sound. This concept of using continuous electromagnetic waves to transmit sound was a founding principle of radio conceived by Reginald Fessenden, a Canadian inventor.

Fessenden is a relatively obscure name to most despite the fact that his breakthrough discoveries affected US military operations, audio and visual entertainment, and wireless communications—all major American industries today. In his youth he dreamed of working for Thomas Edison and pursued that dream until he was hired as an assistant tester for Edison

Above: American radio pioneer, Reginald Fessenden, who worked with Thomas Edison.

Machine works in 1886. He was later promoted to junior technician, working directly with his idol.

After a few fruitful years working in the electrical field and as a professor of engineering, Fessenden was contracted by the United States Weather Bureau in 1900 to develop a network of coastal radio stations to transmit weather information. On December 23, 1900 in Rock Point, Maryland, he successfully made the first radio transmission over a distance of about one mile using a high-frequency spark transmitter. A network was erected between Maryland, Virginia, and North Carolina, but by 1902 Fessenden had resigned from the Weather Bureau due to conflicts with his manager over patent rights.

He quickly partnered with financiers Thomas H. Given and Hay Walker, Jr. to form National Electric Signaling Company (NESCO) in Brant Rock, Massachusetts. Fessenden's first order of business was developing a synchronous rotary spark-gap transmitter, which would produce clear signals with an almost musical tone that could be audible through atmospheric noise and interference. He completed it in December 1905, and in January 1906, he used this technology to exchange Morse code messages between Brant Rock and a station he had established in Machrihanish, Scotland. This was the first successful two-way transatlantic radio transmission.

Fessenden theorized that simply applying a high-frequency current to an antenna could produce a wireless signal. At first, his theory was dismissed as impossible; it was unanimously believed that an electric spark was necessary to produce radio waves. However, waves produced this way could not maintain signals strong enough to produce good quality audio. As a result, Fessenden commissioned General Electric (GE) to devise a high-frequency alternator that could operate at 50 to 100 kilohertz. For two years,

GE failed to produce a machine that met his specifications, so in 1905 he decided to built one himself.

By November 1906, Fessenden had his high-frequency alternator, and continuous-wave transmission was finally possible for his wireless radiotelephone. He added a carbon microphone to the transmission line to achieve amplitude modulation (AM)—a type of radio signal in which the strength of a radio wave is varied in order to carry information from a transmitter to a receiver. On December 21, 1906, he demonstrated his technology to engineers from AT&T, Bell, and Western Electric, among others. That day, he successfully broadcast voice and music from Brant Rock to Plymouth, Massachusetts. While many were impressed by his invention, it was not yet sophisticated enough to market to consumers. It wasn't until the invention of vacuum tubes in 1910 that the system evolved into the technology that we use today.

NESCO did not reap financial rewards from Fessenden's innovative work. This caused tension among the partners, leading them to dismiss Fessenden—whose

Left: Radio station on the passenger steamship *Titanic*, 1912.

A WORD ABOUT

Some historical records claim that Reginald Fessenden broadcasted a brief radio program—the first of its kind—on Christmas Eve of 1906. On this broadcast, he played a phonograph record, played the violin, sang, and read a Biblical passage over his radiotelephone to listeners aboard ships from the US Navy and United Fruit Company. However, there is little reliable evidence of this, indicating that it may have been embellished for the history books.

temperament was prone to volatility—in 1911. He would take his genius elsewhere.

Deciding to abandon his work with radiotelephony, Fessenden became an early pioneer of sonar technology at Submarine Signal Company in Boston. Disturbed by the fate of the infamous maiden voyage of the *Titanic*, he used the principles of sonic frequency and echo sounding to locate potentially hazardous objects in the water. In 1912 he began to develop an acoustical echo-ranging device that he called the Fessenden oscillator. The device sent a burst of sound from an underwater transducer. When the sound wave came into contact with an object, an echo was generated; the distance to the object was determined by measuring the length of time between the initial sound and the return of its echo. In a test on the USS *Miami* in September 1914, the Fessenden oscillator detected icebergs as far as 2½ miles out. In another test on the USS *Aylwin*, the device picked up signals from a moving submarine up to 5½ miles away. The device could also transmit Morse code messages through the water. The technology was used on submarines in World War I.

Reginald Fessenden's contributions are relatively well-kept secrets. Although not generally recognized as an inventor of the television, he does hold patents dated 1936 for a "television system" and "television apparatus." His advancements laid a foundation for an entertainment industry that would redefine American lifestyle and culture in the 20th century.

Orville and Wilbur Wright

"If birds can glide for long periods of time, then...why can't I?" wondered Orville Wright. In 1899 he and his brother, Wilbur, began studying all aeronautical research available from the Smithsonian. Previously, the Wright brothers' predecessors had only experimented with gliders. Sir George Cayley, an English engineer, was the first to root aeronautical research in the scientific method, and the Wright brothers followed his precedent, eventually taking it further than anyone could imagine.

Through their experiments, the Wright brothers developed wing-warping as a means of aerodynamically controlling lateral balance. They chose Kitty Hawk, North Carolina, as their testing site and tested three models of the Wright Glider between 1900 and 1902. A year later Orville and Wilbur built the Wright Flyer, a model with a 12-horsepower engine, propulsion system, and flight controls. On December 17, 1903, Wilbur flew the Wright Flyer 852 feet in 59 seconds. Modern aviation was born.

PHILO TAYLOR FARNSWORTH & AMERICA'S FAVORITE PASTIME

One of the most beloved inventions of our time was developed by a Mormon boy who lived in a log cabin and rode on horseback to school. It wasn't until the age of 12 that Philo Farnsworth would experience a certain phenomenon: electricity. In 1920, 14-year-old Philo drafted an idea for a vacuum tube that would later become the image dissector.

Six years later, he decided to manifest his vision of an electronic television. He started by learning about electrochemistry, radio electronics, and glass blowing—in order to create the vacuum tubes. On September 7, 1927, Farnsworth painted a black line onto a glass slide and inserted it between the image dissector and a carbon arc lamp, and the image of the line was successfully transmitted to the receiver. The electronic television was born, and households were never the same again.

Years of legal battles with RCA continually favored Farnsworth. Eventually RCA did market and sell the first commercial televisions, but only after paying Farnsworth more than $1 million in royalties and licensing.

THE TRANSISTOR BRINGS TRANSITION

ONE CHIP MAKES HUGE CHANGE

By the 1940s, new technologies were sweeping the country, and Americans were regularly experiencing life-altering inventions. Still, the transistor inspired an unusual level of change, acting as a gateway into the digital age.

Most people take this tiny piece of machinery for granted, but just about every electronic device we touch today is powered by the transistor. A game-changer of 20th-century technology, the transistor put information, power, convenience, and recreation

Silicon is a chemical element with the symbol Si and atomic number 14. As a semiconductor material, it possesses electrical properties that can be chemically altered to conduct electricity in different ways and for different functions.

at our fingertips. by making it easier to build processors to run calculations for electronic devices.

In simple terms, a transistor is a switch. When it is turned on or off, electrical current either flows or stops. Before the transistor, vacuum tubes performed this function, but they were fragile and required tons of power. Mervin Kelly, the director of research at Bell Labs, realized that the company needed to create a more efficient alternative to the triode (an electronic amplifying vacuum tube) in order to expand its telephone business. In 1947, he assembled a team to tackle this mission, headed by physicists William Shockley, John Bardeen, and Walter Brattain. They discovered that touching two gold points to a slab of germanium, a metalloid element and common semiconductor material, produced a signal with an output power greater than the input power, thus acting as an amplifier. Thus the palm-sized point-contact transistor was born. Over the next few years Shockley would develop it into the junction transistor, which would become the archetypal design of the device for decades to come.

Music finally became portable when transistor radios hit the market in the mid-1950s. They may not have been MP3 players, but the compact, portable AM radios would have been unthinkable in the days of hot, heavy vacuum tubes. By the late 1950s, engineers were already thinking *smaller*, in the literal sense. In a complex circuit, if the components are too large or the interconnecting wires too long, electric signals travel more slowly, leading to slower processing times that render a devices less effective. In the 1950s, individual circuits were still being constructed by hand, with components soldered in place and connected with metal wires—an impossibly tedious task. Engineers began experiencing a phenomenon they called the "tyranny of numbers," meaning they could not develop devices with increased performance because too many components were involved.

TEX TECH

The first hand-held calculator, made possible by transistors, is developed in 1967 by Texas Instruments with Jack Kilby at the helm of the project. The prototype was named "Cal-Tech."

In 1958 an electrical engineer at Texas Instruments named Jack Kilby solved this problem. He discovered that a circuit's components (transistor, capacitor, and resistor) could be etched into the same block of semiconductor material (he used germanium) and connected by pieces of fine gold, eliminating the need for individual components and complex wiring. This was the integrated circuit—commonly known as a chip. A few months later, an entrepreneur and engineer named Robert Noyce made improvements to the design. He replaced germanium with silicon, and instead of gold wires, he layered thin aluminum on top of the block and sliced out the excess to connect the components. This made it easier to produce smaller circuits en masse.

The first generation of computers had been powered by vacuum tubes. In the 1950s the transistor ushered in the second generation of computers, which were smaller, faster, cheaper, and more energy-efficient. But it was the integrated circuit that truly changed everything, giving rise to the third generation. These computers would run user-friendly operating systems and require keyboards and monitors. Just as the transistor was the building block for the processor, so the integrated circuit was for the microprocessor. Nowadays, integrated circuits can hold several billion transistors and electrical components in an area smaller than a fingernail. Laptops, smartphones, and tablets—all of these portable devices are powered by tiny transistors within an integrated circuit. It was clear by the late 1950s that the United States was turning into a futuristic new world—technologically and economically. Mass production made new technology affordable and accessible to consumers, and conforming to a high-tech lifestyle started to become the new norm.

William Shockley, John Bardeen, and Walter Brattain were awarded the Nobel Prize in Physics in 1956. Shockley left Bell Labs and founded Shockley Semiconductor in Mountain View, California. That was the beginning of Silicon Valley and everything it would come to represent.

MAKING MORE WELCOMING THE DIGITAL AGE

GAINING ACCESS TO A WHOLE NEW WORLD

The late 20th century was marked by the burgeoning of computer technology. The United States was experiencing an industrial revolution that had analog, mechanical, and electrical technologies evolving into digital forms. Devices were shrinking and so were their components, which were being replaced by something more intangible—programming languages with the power to control machines big and small.

The third generation of computers emerged around 1963. These were large, corporate, mainframe computers, and some utilized a time-sharing system that allowed multiple people to use one mainframe computer processor simultaneously. Personal computers were still non-existent, and minicomputers were an interim technology designed for process control and telephone switching. As a result, computing no longer had to be confined to a central facility; individual departments within a large organization could split up computing work based on different needs. Digital Equipment Corporation was the pioneering company for minicomputers. They developed the

1980s elementary school boys playing an early Radio Shack TRS-80 computer game.

PDP series—short for Programmed Data Processor—in the 1960s. These devices were the first to introduce interactive computing, which allowed users to receive immediate direct feedback through line printers and cathode ray tubes used as monitors.

Hewlett-Packard (HP) also tried its hand at minicomputers when the market was hot. At the time, HP primarily made scientific instruments. In 1966 it introduced the HP 2116A. The device offered the same functions as minicomputers but had a stronger emphasis on scientific instrumentation. One series of HP interface cards could connect the 2116A to instruments such as counters, nuclear scalers, electronic thermometers, digital voltmeters, AC/Ohms converters, data amplifiers, and input scanners that could be used for systems of automated testing. Another series of cards could connect it to certain input/output

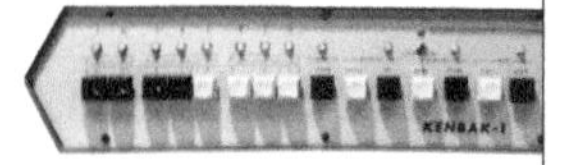

THE FIRST PC

Some regard the Kenbak-1 as the first personal computer. Designed by John Blankenbaker in 1970, it featured integrated circuits on a single circuit board, switches and lights for input and output, and 256 bytes of memory. Only 40 units were sold before the Kenbak Corporation shut down. Not that it was much of a corporation to begin with—it consisted mainly of Blankenbaker working independently out of his garage.

devices like magnetic tape recorders, teletypewriters, paper tape readers and punches, and modems. Although intended for instrumentation, the first HP computer was commercially successful, but mainly for business applications. The demand for personal computers was kindling.

The Intel 8080 8-bit microprocessor was a game changer in 1974. Previously, computers were produced in their entirety by individual manufacturers. However, the 8080 was designed to be integrated into other products and applications. A microprocessor—the key to the fourth generation of computers—is an integrated circuit that contains all of the functions of a computer's central processing unit (CPU). The Intel 8080 offered the nucleus of a computer in a single chip. Hobbyists would be among the first to take advantage of this.

A company called Micro Instrumentation and Telemetry Systems (MITS) developed one of the first microcomputers, the Altair 8800, using the Intel 8080.

* **Nerd alert!** Steve Wozniak was in the Homebrew Computer Club, an early computer hobbyist group in Silicon Valley that met in the 1970s and '80s.

In order to sell units, the struggling company worked to get its computer on the cover of *Popular Electronics* magazine in January 1975. The magazine offered Altair 8800 kits for sale to readers who wanted to build their own computers. MITS needed 200 orders to stay afloat and received a staggering 2,000. Among the people intrigued by the cover story were a young Bill Gates and Paul Allen, who joined MITS to write software called Altair BASIC. At roughly the same time, Gates and Allen's soon-to-be rivals, Steve Wozniak and Steve Jobs, were designing and hand-building the Apple I. It was a fully assembled circuit board with more than 60 chips that could be combined with a keyboard and a television set to function as a computer.

In the 1970s gaming was gaining mainstream popularity. In 1972 a new company called Atari, Inc. launched what would become an iconic video arcade game: *Pong*. By 1977, it would release the Atari 2600, one of the first consoles to put home video gaming on the map. Both arcade and home video games had become billion-dollar industries by the 1980s. However, video gaming experienced a "crash" in 1983 when consumers began turning to the newer, shinier home computer for entertainment. Before long, PC gaming would develop, and computer giant Microsoft would launch the Xbox.

Meanwhile, the Internet was emerging, adding a new dimension to the personal computer and gaming technology that was already infiltrating the American lifestyle. The first incarnation in 1969 was the Advanced Research Projects Agency Network, or ARPANET. It started as a network of four geographically separated computers set up to protect military information. As the network grew, it became more difficult for the computers to connect. By the end of the 1970s, Internet pioneer Vincent Cerf had developed transmission control protocol (TCP) to enable computers throughout mini-networks around the world to connect and communicate. Scientists and researchers made great use of the network. Consequently, the growth of the network also compromised security, and ARPANET dissolved and assumed the form of what we know now as the Internet.

Computer programmer Ray Tomlinson had already written the first basic email software (for ARPANET in 1971), which became a valuable network application. Soon, the Internet became more than a networking tool; it was a "web" of information. In 1992, students at the University of Illinois developed Mosaic—later renamed Netscape—a user-friendly browser that could display pictures, text, and clickable links. The first wireless Internet network was built at Carnegie Mellon University in 1994, and connecting to the World Wide Web was never the same again.

Perhaps the biggest success of this digital age has been the smartphone. With it, we have instant access to just about anything, from anywhere. It started with IBM's Simon, the first cellular phone with features for managing personal information, and then the PalmPilot came on the scene offering more advanced capabilities. Now smartphones control our lives. The digital age has interwoven computer technology with everyday life.

THE TECH LIST
LASER FACTS

1 **LASER** is an acronym for "light amplification by stimulated emission of radiation."

2 **Focusing light** on a concentrated spot over a distance is the main feature of a laser.

3 **Theodore H. Maiman** beat out major research groups in developing a functioning laser in 1960.

4 **Maiman's design** used a helical flash lamp to light up a synthetic pink ruby crystal.

5 **Laser scanning** was implemented for the first time at the Marsh supermarket in Troy, Ohio in 1974.

6 **Laser-guided munitions** were developed for the US Army in the 1960s.

LAYING NEW FOUNDATION
THE PIONEERS OF DIGITAL TECHNOLOGY

INTEL, IBM, AND SUN MICROSYSTEMS

A young visionary, Bob Noyce—the co-inventor of the integrated circuit—walked out on Shockley Semiconductor to cofound Fairchild Semiconductor with his colleague Gordon Moore in 1957. Moore hired a researcher, Andy Grove, to complete the trifecta that would revolutionize the personal computer industry. When Noyce and Moore left Fairchild in 1968

Left: Fairchild Semiconductor division head Robert Noyce in his office with diagrams of semiconductors and microchips.

to start another company, Grove tagged along. He would be the first "hire" at Intel Corporation.

Intel's company culture was casual from the start; Noyce shunned the stifling, bourgeois ideal of corporate life in favor of an inspired, creative environment, setting the blueprint for the now-ubiquitous model of Silicon Valley tech culture. With Noyce's charismatic confidence, Moore's technological virtuosity, and Grove's natural business savvy, Intel created a niche in the computer industry and thrived for decades within it.

In the beginning, Intel set out to do one thing: make practical semiconductor memory hardware. But their focus shifted when Intel engineer Ted Hoff invented the 4004 microprocessor in 1971. They realized that Hoff's chip could be plugged into other applications and machines and perform as a "brain." In 1972 Intel topped the 4004 with the 8008, which could process the same amount of information twice as fast. By the release of the 8080 in 1974, microprocessors were being used in supermarket scales, traffic-light sensors, medical instruments, gasoline pumps, and pinball games, just to name a few.

The opportunity of a lifetime came along in 1981 when IBM became interested in Intel's microprocessor for the development of IBM's first personal computer. Over several secret meetings, during which a black curtain separated Intel engineers from IBM engineers and their PC prototype, IBM was convinced to use Intel's 8088 processor for IBM's 5150 model scheduled to launch later that year. When Intel introduced its 286 chip, with 134,000 transistors, IBM chose it for their 1984 PC model. By the 1990s, with Andy Grove at the helm as CEO, the Intel microprocessor had become iconic to the PC movement. The 1993 Pentium processor was about 1,500 times faster than the original 4004.

IN THE MID-1970s, SOMEONE CAME TO ME WITH AN IDEA FOR WHAT WAS BASICALLY THE PC . . . I ASKED, **'WHAT'S IT GOOD FOR?'** —GORDON MOORE

Computing-Tabulating-Recording Company (CRT), the predecessor to IBM, was founded by Charles Flint in 1911. It became International Business Machines Corporation (IBM) in 1924 under the presidency of Thomas Watson, Sr., but it was his son and successor, Thomas Watson, Jr., who ushered IBM into a golden age starting in 1952. After decades of research and development and technological business solutions, IBM produced its first large-scale electronic computer, the 701 Electronic Data Processing Machinein 1953, and a future tech giant was born. The IBM 1401 data processing system in 1959 became the best-selling computer of the time. By 1964, the company had developed System/360, a family of computers with compatible software and hardware—the first of its kind. When in-house researcher Robert Dennard developed DRAM memory—memory cells that store information as electronic charges in electrical circuits—in 1966, IBM set the stage for personal computing. It would be over a decade, however, before that would come to be.

Of IBM's contributions, perhaps the most popular was its introduction of its first PC. In 1980, IBM commissioned computer programmer Bill Gates to develop an operating system. Powered by the Intel 8088 microprocessor, the IBM 5150 (also known simply as PC) hit the market in 1981. It had 16KB of RAM, no disk drives, and a few applications, including a spreadsheet, VisiCalc, and a word processor, EasyWriter. The personal computer was no longer just in the hands of hobbyists—laypeople were finally incorporating the technology into their homes. *Time* magazine's honor of "Man of the Year" in 1982 went to IBM's PC—anointed "Machine of the Year."

Sun Microsystems had its heyday as well, but its history is more brief. Founded by Stanford graduate students Vinod Khosla, Andy Bechtolsheim, and Scott McNealy in 1982, the company launched a powerful operating system called UNIX. Bill Joy, one of the leading UNIX developers at the time and the "Edison" of the Internet era, later joined as a co-leader. Sun specialized in technical workstations, and within the industry, its computers were the gold standard for serious technical computing. In 1995 it introduced the object-oriented programming language Java.

Sun thrived during the "dot-com bubble" around 1997 when the widespread growth of Internet users spurred investors to finance tech start-ups, which they assumed would become profitable. When many of those companies failed, the bubble burst in 2000 and triggered an economic recession. Though Sun had been at the pinnacle of success, it didn't have the business acumen of other companies like Microsoft and failed to recuperate from the crash. In 2010 Sun was acquired by Oracle Corporation.

INDIA'S TECH TYCOON

GETTING RICH AND GIVING BACK

In 1945, a hydrogenated cooking fat company called Western India Vegetable Products Limited was launched in India. By 1981, with Azim Premji at the helm, the company had changed its name to Wipro Limited and entered the information technology sector. In 2015 Wipro boasted over $7 billion in annual revenue. Premji believes in social responsibility, employing ethical and ecological business practices and engaging with societal issues in India. He has dedicated more than $4 billion of his fortune to the country's public education system, and was the first Indian to join The Giving Pledge, a campaign that asks billionaires to give most of their wealth to charity.

THE WORLD'S MOST POWERFUL PEOPLE

GETTING TO KNOW AMERICA'S TECH BILLIONAIRES

Every year, *Forbes* magazine publishes a list of the world's most powerful people. In the United States, some of these success stories started with nothing more than an idea and the will to make it work by any means necessary.

Michael Dell was a pre-med freshman at the University of Texas at Austin in 1984 when he decided to create his own business offering upgrade kits for personal computers. He founded PC's Limited—later known as Dell—with just $1,000 and worked out of his dorm room. A year later, Dell built its first computer system, the Turbo PC. By 1988, the company had made its first public offering, turning Dell a multimillionaire by age 23. After

Google's cofounders Larry Page and Sergey Brin (right); Amazon founder and CEO Jeff Bezos; and Facebook founder and CEO Mark Zuckerberg, were also included on *Forbes*' 2014 list of The World's Most Powerful People.

sales doubled for the third year in a row in 1991, he became the youngest person ever on the Fortune 500 list in 1992. Sales continued to soar in the United States as well as internationally.

Because its business model focused on supply chains to sell existing technology at lower prices, Dell lacked the innovative power to branch out into the lucrative new mobile technology market. In 2004 Michael Dell stepped down as CEO but remained chairman of the board. Three years later the board was begging for him to return to the helm, which he did. After a battle with investors—with activist shareholder Carl Icahn being the most stubborn—Dell privatized the corporation in 2013 with hopes of revitalizing its business. Now the company focuses on data security, networking, and cloud computing.

If there was ever a "bad boy" of Silicon Valley, it would be Larry Ellison. The founder of a powerful business software, Oracle Corporation, Ellison is known for his love of fast yachts and rock and roll, as well as his tendency to attract controversy. After building databases for the CIA, Ellison saw a need for relational databases that would recognize connections between stored pieces of

DID YOU KNOW

Fluid Concepts and Creative Analogies: Computer Models of the Fundamental Mechanisms of Thought by Douglas Hofstadter was the first book sold on Amazon. Although the company started as an online bookstore, today Amazon's diversified market has merchandise ranging from electronics and digital media, to apparel, jewelry, groceries, and more.

information. At that time in 1977, only trained professionals could operate mainframe computers for inputting and outputting data. Software developers still wrote code using pen and paper. With American businessmen Bob Miner and Ed Oates, Ellison started Software Development Laboratories, which later became Oracle Corporation. By 1978, the three launched their namesake software, Oracle Version 1, with each subsequent version revolutionizing business computing.

By 1995 they were strategizing Internet compatibility for all of its products, ahead of customer demand. Four years later they achieved it, just in time to survive the dot-com bubble collapse in 2000. When its competitor Microsoft became involved in an antitrust lawsuit that year, Ellison took the liberty of hiring investigators to dig through the trash of the Independence Institute and the National Taxpayers Union in order to expose their ties to Microsoft. Ellison was not sorry at all for his stunt, having revealed that the organizations were, in fact, being funded by Microsoft to advocate for them during the suit.

Despite Ellison's vision and tenacity, Oracle has not been able to bypass Microsoft as the number one software company. Microsoft cofounder Gates is legendary for his brilliance, competitive spirit, and aggressiveness in business. His software programming career began early in high school when he and his classmate Paul Allen started dabbling in computers at the computer lab, sometimes getting into trouble for their antics. When Gates went to study law at Harvard University, he couldn't shake his obsession with computers. He dropped out of school and went to Albuquerque, New Mexico, with Allen to work with MITS. It was there that Gates grew frustrated with the open-source sharing of software that was popular among computer hobbyists;

Microsoft Corporation CEO, Bill Gates, poses in his office with a copy of Microsoft Excel (1988).

he believed that it was counterproductive to the development of newer, better software.

Gates and Allen started Microsoft in 1975. Five years later, Gates's friend from Harvard, Steve Ballmer, came on board as business manager. The young entrepreneurs managed to convince IBM to let them create the operating system for IBM's first PC. The result was MS-DOS. Amid all of this, Allen was diagnosed with non-Hodgkin's lymphoma in 1982. Although he went into remission a year later, he resigned from his position as Microsoft's executive vice president.

In 1985 Microsoft released its new operating system, Windows 1.0, which allowed users to navigate the system with a mouse instead of confusing key commands. By the release of Windows 95 in 1995, Microsoft had become synonymous with the word "PC." This period would burnish the company's cleverness for marketing strategies; Microsoft possessed an early understanding of branding (something that would become ubiquitous in the impending cultural renaissance of the Internet). The same year, Microsoft released its browser, Internet Explorer, before the dot-com bubble burst. Gates stepped down as president of Microsoft in 2000, passing the torch to the exuberant Ballmer, who led the company on a departure from its PC heritage by focusing on other products such as the Xbox video game consoles and the telecommunication software, Skype.

Jeff Bezos named his company Amazon, after the largest river in the world, with dreams that it would become the biggest store in the world.

By the dawn of the new millennium, the Internet was providing a whole new canvas for innovation. Computer scientists Sergey Brin and Larry Page were quietly planning their entrance from the sidelines of Silicon Valley. After meeting at Stanford in 1995—and being initially irked by one another—the two decided to collaborate on a search engine called Backrub, which they later renamed Google. The name comes from a play on the word "googol," which is 10 raised to the 100th power, and is meant to evoke the very large amount of information found on the web. By the end of 1998, *PC* magazine was praising Google as the top search engine.

Today, Google is so formidable it hardly needs any explanation; its name has become a verb that means to search for something on the Internet. Brin and Page have managed to optimize the way users explore the World Wide Web. Google's tools influence the way people navigate the modern world. The Google Maps app turns cellphones into GPS systems, and Gmail is one of the most widely used email services in the world. Google's Android operating system has become the top competitor to the iOS operating system used on Apple's wildly popular iPhone and iPad devices. However, Google's rise hasn't been without its share of failures. Its optical head-mounted display that allows users to surf the Internet hands-free, Google Glass, has experienced lukewarm reception, as have some of its other products. But that doesn't stop them from experimenting with bold, almost fanciful ideas, like self-driving cars.

Jeff Bezos named his company Amazon, after the largest river in the world, with dreams that his company would become the biggest store in the world. It's difficult to quantify, but Amazon has come close to being just that—a huge marketplace where consumers can purchase just about anything from books to computers to monster gas grills. When Bezos started his company in 1994 out of his garage in Bellevue, Washington, the servers required so much power that a flip of a switch of a blow dryer would blow a fuse. Bezos originally chose books as his niche because of the demand for literature and the large quantity of titles available around the world. In two months, Amazon was selling books to buyers in all 50 states and more than 45 countries.

Jeff Bezos announcing the new Kindle DX electronic book reader in 2009.

The first few years brought growing pains, however. Bezos—known for being a tough boss—required his employees to work 60-hour workweeks to keep up with the growth of the business. Even so, the company did not turn its first profit until 2001. In 2015 Amazon is the go-to site for purchasing personal and household items. It has attracted customers by helping customers receive items faster and more cheaply. Its crowning jewel—the Kindle—elevated Amazon's mission by digitizing books for customers to purchase, download, and read on a single device.

After a stint as a professional gambler in his youth, Charles Ergen partnered up with his wife, Cantey, and a friend named Jim DeFranco to sell satellite dishes out of the back of a pickup truck in Denver, Colorado. In 1980 they called their business

EchoStar. By 1992 they were able to obtain a direct-broadcast license from the Federal Communications Commission granting them their own geostationary orbital slot. They were incorporated in 1993, switching their name to DISH (Digital Sky Highway) in 1996. They became the first satellite provider to offer two-way high-speed Internet and introduced the digital video recorder to the masses. In an attempt to woo customers, they introduced a feature called AutoHop, which allowed users to skip past television commercials. Networks like NBC, CBS, ABC, and Fox filed a lawsuit arguing that the feature interfered with their livelihood in advertising. The case is still open as of 2015 and DISH remains one of the top satellite providers in the country.

The tale of Twitter is one of great drama, centering around the personality of one of its cofounders, Jack Dorsey, an NYU dropout. While working as a disheveled engineer for Evan Williams—who had previously banked big success for selling his company Blogger to Google—Dorsey came up with the idea to create a micro-blogging platform that would allow users to share up-to-the-second details about their everyday lives. His friend and coworker, Noah Glass, refined the idea to

THE SNOOP

Oracle Chairman Larry Ellison proudly took credit for hiring Investigative Group International to snoop through Microsoft's ties, claiming Oracle was doing its "civic duty." A Chicago Tribune *article quotes Ellison as confirming, "I feel very good about what we did." The same article claims that "The work by IGI allegedly included a $1,200 offer to janitors to get a peek" at the Association for Competitive Technology's trash.*

focus on human connection through this social media network. Glass named it Twitter. They went live in 2006 with the first "tweet" by Dorsey. It was a rough road for Twitter from the beginning, with frequent system outages and difficulties upgrading the site from its prototype. Meanwhile, there was a power struggle between Williams and Glass. But it was Dorsey who secretly gave Williams an ultimatum to drive out his friend Glass, whose emotions were getting in the way of Twitter's mission.

With Glass out, Dorsey assumed the role of CEO, but he became distracted by the Silicon Valley lifestyle, which had him attending San Francisco Giants games with venture capitalists. In 2008 the board ousted him as CEO and made him a silent chair instead. When Dorsey was wooed by Mark Zuckerberg to join Facebook, Dorsey instead took to the media and began promoting himself as the face of Twitter. Twitter's members and the general public soon accepted his story that he was the brains behind Twitter. To an extent, he was, though that was not the complete story. He returned to Twitter as executive chairman in 2011.

ONE BAD APPLE SAVES THE BARREL

THE UNIQUE HISTORY AND LEGACY OF STEVE JOBS

Just as Apple is an icon of technology, so too is its founder, Steve Jobs. A highly contentious man whose life story was marked by brilliance, perseverance, and controversy, Jobs was probably one of the most admired leaders in modern technology. He left a legacy of technology that has influenced the way people live and view the role of technology in their lives.

In the beginning, Paul Jobs taught his adopted son, Steve, how to take things apart and put them back together. This sparked a lifelong interest in technical tinkering which was

In his final year at Apple, Steve Jobs only made a $1 annual salary. He didn't need the money, however, because he held 5.5 million shares of Apple and 138 million shares of Disney. Despite his fortune, Jobs was never known by others to be motivated by money.

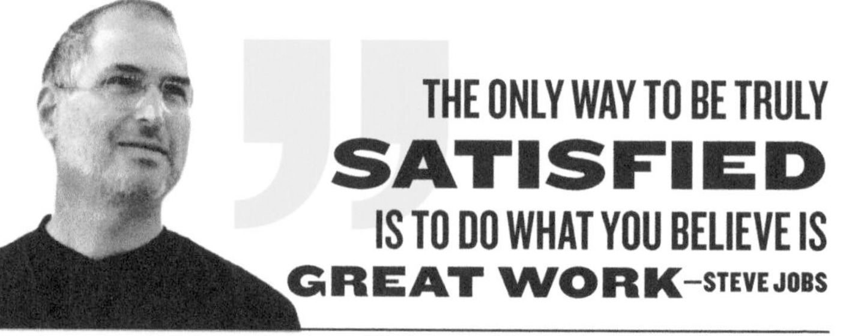

later supplemented with lessons on creativity from college classes and his search for spiritual enlightenment while traveling through India. After returning home in 1975, Jobs met a computer whiz named Steve Wozniak and convinced him to let Jobs sell the computer Wozniak was building, the Apple I. Jobs would handle the marketing. To fund their venture, Jobs sold his VW bus, and Wozniak sold his HP scientific calculator. In 1976 Apple was born.

Apple's history is riddled with successes and failures. Although Apple made $139 million off sales of its second computer, the Apple II, and went public in 1980, rough times were ahead. Apple's subsequent machines were plagued with design flaws, recalls, and overall consumer disappointment. In the meantime IBM was prevailing in the personal computer arena. In 1984 an aggressive ad campaign billed the new Macintosh as the ideal computer for youthful, forward-thinking nonconformists who wanted to avoid the "conformity" of the PC. Unfortunately, the Macintosh's incompatibility with IBM systems alienated a huge customer base. Jobs's vision for his company interfered with the expectations of shareholders and then-CEO John Sculley, so in 1985, Jobs stepped down from his position as chairman of the board.

A year after his departure from Apple, Jobs cofounded the animation studio Pixar. The new venture kept Jobs busy, but his attention eventually returned to Apple, which was struggling by the mid-1990s. In 1997 Jobs came back as CEO, and the golden age for both Apple and Steve Jobs began. Some say it was his domineering, at times cruel managerial style that turned Apple into a paradigm of 21st-century creativity and innovationin both technology and business. Jobs was known for pushing his employees beyond what they thought possible. He could be callous and raw in his criticism, yet very few of his employees ever quit. In fact, most relished the opportunity to work under his command. His ideas were bold and ambitious, and many felt the risk of incurring his wrath was worth the opportunity to participate in something special.

The release of the iMac put Apple back on the map. But perhaps the most important event during this resurgence was the relationship formed between Jobs and Jonathan Ive, whom Jobs affectionately referred to as Jony. An Apple employee since 1992, Ive became the senior vice president of design in 1997. Every significant Apple product to emerge since then—the iMac, MacBook Pro, MacBook Air, iPod, iPhone, iPad—was conceived by Ive over years of inspiration, design, and

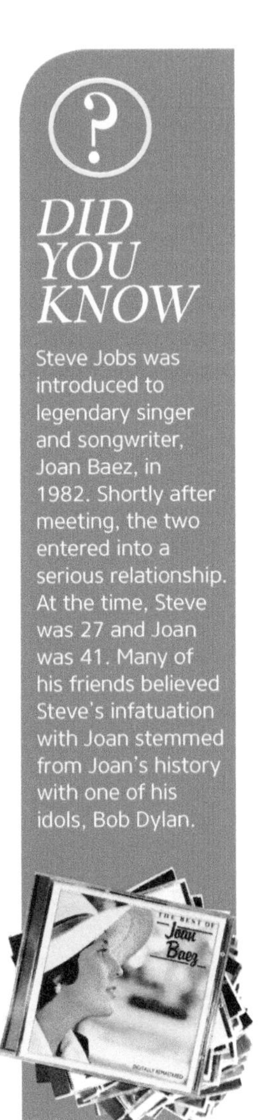

development. He and Jobs shared a sentiment of hating fuss and loving simplicity. This minimalism has been at the core of Apple's design philosophy since Jobs's return—the it-factor that has made the masses follow the brand's every move.

Jobs was fond of closing his keynote speeches with a catchphrase—"One more thing"—before unveiling a new Apple product. These "Stevenotes" often generated a lot of buzz, but it was Ive's designs that propelled Apple's sales to new heights. It was by the grace of Ive that the iPhone was released in 2007. That first year, six million phones were sold. By 2012, more than 100 million iPhones were flying off shelves each year. By the launch of the iPad and Macbook Air, Apple's value had quadrupled. Not everyone is an Apple fan, though. Many people object to the built-in obsolescence and high price points of Apple products. However, Apple has always been clear that goal isn't to create something economical, but to create something that will fit into users' lives in a way they haven't thought of before—and then, to make it better every time.

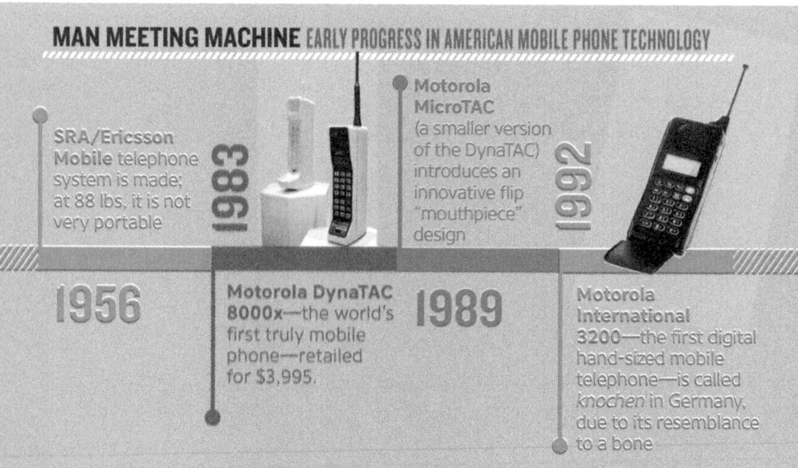

In 2003 Jobs was diagnosed with pancreatic cancer. He passed away in October 2011, leaving Apple in the hands of Ive and the new CEO, Tim Cook. Since Steve Jobs's passing, Apple has yet to release its next game-changing technology.

The Apple Watch might change all of that, however. Conceived in Tim Cook's lab shortly after Jobs' death, the Apple Watch was designed and produced without any involvement from Jobs. Cook and Ive hope it will usher in a new age of wearable technology and uphold the legacy of innovation that Jobs left behind.

uiz

INFILTRATING THE BOYS' CLUB

WOMEN IN TECH

1. In a 2015 study, Google revealed its percentage of female staff to be what?
   A. 42%
   B. 30%
   C. 23%

2. What is the name of IBM's first female CEO?
   A. Virginia M. Rometty
   B. Lisa Davis
   C. Ellen Pao

3. A glamorous "computer geek," Ginni Rometty studied computer science and electrical engineering at which US university?
   A. Brown University
   B. Tulane
   C. Northwestern University

4. What position did Ginni Rometty hold when she first joined IBM in 1981?
   A. Systems Engineer
   B. IT Technician
   C. Lead Manager

5. Renée James joined Intel in 1987 and became president in which year?
   A. 2009
   B. 2013
   C. 2012

6. Who is leading Intel to withdraw from the PC market and explore wearable technology?
   A. Gloria Gains
   B. Sheryl Sandberg
   C. Renée James

7. Early in her career, Renée James served as chief of staff for which former Intel CEO?:
   A. Simon Grass
   B. Andy Grove
   C. Mandy Green

8. Ginni Rometty was instrumental in the development of IBM's artificially intelligent computer, Watson. On which televised game show did Watson compete?
   A. *Who Wants to Be a Millionaire?*
   B. *Jeopardy*
   C. *Family Feud*

9. What is the name of Renée James's $300 million initiative to increase training and recruitment of women and other minority groups in technology?
   A. Women in Technology
   B. More Money, More Women
   C. Diversity in Technology

ANSWERS: 1.) B 2.) A 3.) C 4.) A 5.) B 6.) C 7.) B 8.) B 9.) C

MARK ZUCKERBERG RIGHT OR ROGUE?

FROM ZUCKNET TO TECH MAGNATE

By high school, he had designed Synapse, a precursor to the music software Pandora that could learn listeners' musical tastes. Microsoft offered him $1 million for the software, but the teenager refused to sell. At Harvard, he became the most sought-after programmer in school. The story is familiar to those who have seen the Academy Award-winning film *The Social Network*, which was based on a book by Ben Mezrich, *The Accidental Billionnaires*. The film and book blurred lines between fiction and reality, according to Zuckerberg, but elements of his storied successes and scandals are legendary nonetheless.

FACE TIME

After connecting billions of people around the world with just one website, Mark Zuckerberg was named Time *magazine's Person of the Year in 2010.*

Throughout the course of his career, Zuckerberg has demonstrated sharp business acumen as well as programming prowess. He developed Facebook in one week in 2004, and with his friends Dustin Moskovitz, Chris Hughes, and Eduardo Saverin, he managed the site from a Harvard dorm room. Originally intended as a social network for Harvard students only, Facebook's membership exploded when access was granted to other colleges and universities. By the end of 2004, there were 1 million users. Zuckerberg dropped out of school and moved operations to Palo Alto, California, where companies like MTV, Yahoo, Microsoft, and Viacom all came knocking. In an audacious move, Zuckerberg rejected all of their offers, deciding to stay true to his own unrelenting vision for his company. From the beginning, Zuckerberg has always believed that people should—and will—embrace the notion of making their private lives public.

Zuckerberg shares the boldness and bravado of many tech moguls, and like them, Zuckerberg didn't get where he was without some controversy. While at Harvard, classmates Divya Narendra and Cameron and Tyler Winklevoss commissioned Zuckerberg to help them build a social network called the HarvardConnection, which he agreed to do. But Zuckerberg later abandoned the project in favor of creating his own social network. Narendra and the Winklevosses sued Zuckerberg for allegedly stealing their intellectual property and breaching their agreement. After seven years of legal battles, they eventually reached a $65 million settlement in 2011.

Facebook has been under fire over privacy issues since as early as 2006. That year, the company introduced its "News Feed" feature, giving members the ability to monitor their friends' Facebook activity. Since then, Facebook has become notorious for using restrictive language in their privacy policy and frequently updating privacy options, often causing members' profiles to default to more public settings until users catch wind of the change and manually reset their preferences. As of 2007, search engines were able to index Facebook profile pages, making them searchable through the web.

Facebook has also been accused of data mining—collecting user data from their servers and analyzing it for internal and external use. The "Connection" feature added in 2011 encourages users to click "Like" buttons on both Facebook and external websites, publicizing their feelings about certain products or services. Rolled out at the same time, the "Instant Personalization" feature shares user information that Facebook classifies as public (i.e., name, location, list of friends, Likes) to affiliated websites and advertisers, giving them insight into who a user is and what their preferences might be. Facebook has also been accused of using hundreds of millions of users' profile activity for psychological research. In 2014, the Federal Trade Commission publicized the complaints against Facebook concerning privacy. The FTC has required Facebook to protect user privacy under a 20-year consent decree.

The global surveillance disclosures leaked by American whistleblower Edward Snowden 2013 revealed that the PRISM surveillance program initiated by the National Security Agency had obtained access to Facebook's servers. Zuckerberg denied that any such request was made by government agencies or that Facebook granted access.

With more than 1 billion current users, Facebook is going strong, although Zuckerberg is continually faced with the challenge of adapting his business model for sustained profitability. But because Facebook possesses the largest database of personal information about the world's population, and will likely have little trouble figuring out a way to use this resource to his advantage.

THE TOP 10 SEARCH ENGINES YOU'VE NEVER HEARD OF

1 **Wow.com** is owned by AOL and has 100 million unique visitors each month.

2 **WebCrawler.com** blends the top search results from Google and Yahoo!

3 **MyWebSearch.com** often appears as a tool bar, its search engine is powered by Google.

4 **Blackle.com** keeps the screen black to convserve energy.

5 **Info.com** aggregates results from other search engines.

6 **Dogpile.com** is powered by Metasearch technology.

7 **Infospace.com** helps businesses monetize through searches.

8 **DuckDuckGo.com** doesn't track users' activity, protecting their privacy.

9 **Yippy.com** was created by Carnegie Mellon University researchers.

10 **Alhea.com** includes sponsored advertising and algorithmic search results.

HOW GOOGLE CHANGED THE GAME

THE MAD SCIENTISTS OF THE MILLENNIAL GENERATION

Larry Page was considering Stanford for graduate school in 1995 when he met an outspoken grad student named Sergey Brin for a tour. Although they spent the day matching wits, an unusual friendship was forged. From the beginning, Page was the introverted mastermind and Brin was the extrovert with a knack for strategy, branding, and networking. Together they raised $1 million from friends and family for a start-up called "Backrub." It was 1996, and Page had dreamed up an algorithm for a search engine that he believed would one day enable users to access all the information in the world quickly and easily. The name "Backrub" didn't work, but the product—renamed Google—did.

They started to see success by 1998 and incorporated their business. A year later, Page came up with a way to optimize their search engine. Servers are typically hosted in rooms owned by third-party warehouses, which charge by the square foot. Page realized if he shrank the physical size of the servers, he could fit more of them into a room. He stripped them of all unnecessary components and crammed 1,500 slim servers in the same amount of space that their competitors used to host 50 servers. The extra servers allowed Google to run searches exponentially faster than their rivals.

Business began to get serious, especially after they secured $25 million in funding from bigwig venture capitalist firms Kleiner Perkins and Sequoia Capital. There was a catch, however—they wanted Page to step down as CEO and hire a professional who could essentially babysit the two 20-something founders. Page—a notorious control freak—resisted. He opened to the idea, however, after taking meetings with a few successful tech CEOs: Jeff Bezos, Andy Grove, and Steve Jobs.

In 2001 Eric Schmidt came on board as chairman, and eventually, CEO. Page was named president of products, and Brin the president of technology. By that time, Google had launched in several languages: French, German, Italian, Swedish, Finnish, Spanish, Portuguese, Dutch, Norwegian, Danish, Chinese, Japanese, and Korean.

In 2000, Google played the first of its annual April Fools' Day hoaxes. It announced the MentalPlex—a new technology that would allow Google to read users' minds as they visualize their search topic.

Twelve-year-old Larry Page was brought to tears after reading Nikola Tesla's biography in 1985. Tesla was a Serbian American inventor and electrical engineer most known for his contributions to the modern-day electricity supply system. Though an idealistic and brilliant genius, Tesla was a terrible businessman. Thomas Edison swindled him and investors gave up on him. In the end, Tesla died alone and in debt.

The Google toolbar was introduced, allowing users to search the Internet without going to Google's homepage, and Google Images gave users access to 250 million searchable images. Things were moving fast, but Page grew weary of the hierarchy that put project managers between him and his engineers. He publicly fired all project managers at a meeting in 2011, which outraged everyone, including the engineers, who actually preferred to work under managers. Schmidt diffused the situation and brought in a VP of product management, Jonathan Rosenberg. Rosenberg would find it challenging to please Page, who despised bureaucracy, until engineer Marissa Mayer suggested that they hire computer science graduates to fill the roles of project managers. The move seemed to put Page at ease.

Under Schmidt's guidance, Google made more progress. Gmail launched in 2004, and the company went public that same year. After that came Google Maps and Google Earth in 2005, and the YouTube acquisition in 2006. With Page settling into his role and allowing Schmidt to do his job, Page took on an acquisition of his own—he bought a start-up called Android for $50 million. It was chump change for Google, so Schmidt didn't concern himself with it. Page worked with Android's CEO Andy Rubin on what would be Google's next huge venture.

In 2007 the Android operating system was announced, and the following year, Google released the G1 phone to compete against the Apple iPhone. In addition, Google made the Android operating system for free for phone manufacturers. Soon enough, companies like Samsung and Motorola were jumping at the chance to run the Android system on their phones, and service providers could compete better against AT&T, which was the exclusive iPhone provider at the time. In 2010 Google made its next move and

Google headquarters in Mountain View.

launched the Nexus line of Android smartphones and tablets. That year, Android surpassed Apple's iOS as the top-selling operating system for digital devices.

Google's core business as a search engine flourished, becoming the Internet's most successful advertising business. But Page worried that the company had lost some of its ambition. He wanted to get back to creating products that people felt they couldn't live without. In 2011 a more mature Page stepped up and reclaimed his seat as CEO, giving Schmidt the position of executive chairman. Page decided to focus on more wild innovations like self-driving cars and artificial intelligence.

In 2012 "Project Glass" made its first appearance. A pet project of Sergey Brin, it was initially referred to as "Google X"—an undisclosed project that took place in a secret laboratory where a small group of engineers toiled away at something no one else in the company was allowed to know about. By the time they had a working prototype, Brin was excited to test Google Glass in public. Instead of retailing the first version—still just a model in development—Google created a program called Glass Explorers, which allowed a group of tech enthusiasts and journalists to pay $1,500 for the privilege of testing out the technology. The glasses generated a great deal of buzz. Movie stars flaunted them, and models wore them on the runway at a Diane von Furstenberg fashion show—with Brin making a special appearance. Users seemed more perplexed than anything else by the technology, and privacy concerns over the built-in camera got the device banned from certain public places. Google announced in 2015 that it was shuttering Google Glass and sending the project to Tony Fadell, a former Apple executive best known for leading the team that created the iPod.

Among Google's other new innovations are an ultra-high-speed Internet connection service called Google Fiber, and Calico, a subsidiary company that will focus on health issues related to aging and age-related diseases. Google has also acquired Boston Dynamics, a company that creates humanoid and animatronic robots. Given Google's success in developing innovative new products, it seems that the mad scientists of the millennial generation are not so mad after all.

FEMALES UP THEIR GAME

WOMEN IN TECH

WOMEN IN THE C-SUITE

Times are changing, and more and more women are gaining something that has long been out of their reach in the tech industry: power. Two trailblazers who have led the way are Marissa Mayer and Sheryl Sandberg. When Mayer graduated from Stanford in 1999, she received 14 job offers. She chose a fledgling company called Google. At the age of 24, she was Google's first female engineer and their 20th hire. Mayer's intense work ethic and intelligence helped her rise up the ranks. She became a spokesperson for Google and played many key roles, working

Above: Ursula Burns, President of Xerox Corporation, 2009.

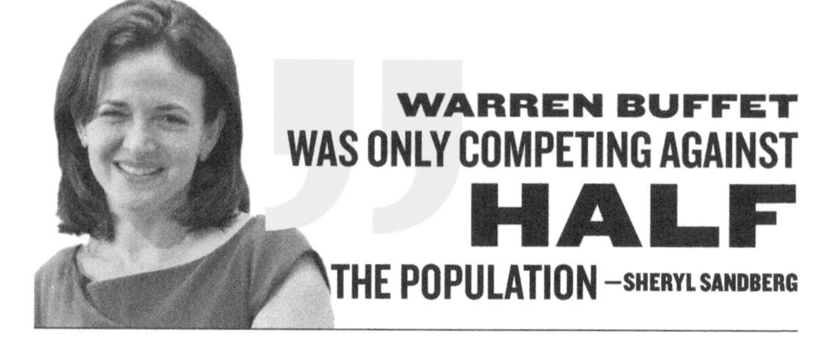

as a software engineer and product manager, overseeing the minimalist design layout of the homepage, and heading company acquisitions. But with all of the immense talent working at Google, Mayer eventually hit a ceiling. So she made a bold move and accepted a position as president and CEO of Yahoo! in 2012.

Sheryl Sandberg not only embodies what it means to be a high-powered executive woman—she literally wrote the book on it, publishing *Lean In: Women, Work, and the Will To Lead* in 2013. She joined Google in 2001 as VP of global online sales and operations. A chance encounter with Mark Zuckerberg in 2007 resulted in the Facebook founder spending weeks courting Sandberg for a new COO position at his company, back when they barely had any revenue. He was convinced that she was his ideal second-in-command. Up for the new challenge, she turned down a CFO position at Google and made the jump to Facebook by 2008. It took just two years for Sandberg to help the company turn a profit. Her management skills are highly regarded, and her ability to guide companies through stages of rapid growth is legendary. In addition, many credit her remarkable partnership with Zuckerberg as a key reason for Facebook's current success; the two are among the most powerful business partners in Silicon Valley. Sandberg has also used her talent for building relationships outside of Facebook, too, by hosting events that help women network and enter leadership positions.

The truth is, women only represent 24 percent of the science, technology, engineering, and math workforce; as a result, men usually claim executive positions in the tech industry. In addition, many executives have the advantage of middle-class backgrounds to help them succeed. Xerox CEO Ursula Burns was not one of them. Burns grew up in a Lower East Side Manhattan housing project. With the support of her single mother, Burns followed her engineering aspirations and won a scholarship to the Brooklyn Polytechnic Institute. As a black woman, she stuck out in a sea of white male classmates. She earned a degree in mechanical engineering in 1980 and then took a summer internship with Xerox while pursuing a master's at Columbia. The president of marketing and customer operations chose Burns to be his executive assistant in 1989, only to lose her two years later to the CEO, Paul Allaire. In 1997 Burns was named VP for worldwide manufacturing, leading the company into color copying. When Xerox fell on hard economic times in 2000, Burns almost left the company,

FIRST LADY

In 1809, Mary Dixon Kies is the first woman to be granted a US patent. Her invention made it possible to weave straw with silk and thread. The Patent Act of 1790 allowed both women and men to protect their inventions.

until her longtime colleague and friend Anne Mulcahy became CEO. Together, they helped save Xerox from going under, restoring it to profitability by 2004. In a historic move, Mulcahy named another woman as her successor in 2009: Ursula Burns.

There is power in the supportive business relationships among women in tech. IBM's senior VP of global business services, Bridget van Kralingen, is often considered the protégé of its CEO, Ginni Rometty. Van Kralingen joined the company in 2004 as managing partner of financial services. Ten years later, she was at the forefront of one of the biggest deals in tech history—a partnership between IBM and Apple. Such a deal might have been unthinkable in the 1980s when the two companies were fierce competitors. But IBM honed in on an aspect of its business that would reposition the future of the company: consulting, systems integration, and application management services. Van Kralingen was at the center of it all. In a deal largely negotiated by her, IBM agreed to help its business clients integrate Apple devices into their workplaces. This came at a time when companies were increasingly interested in boosting productivity and flexibility by helping employees use their personal mobile devices for business purposes. At the same time, Apple was looking to become more enterprise-friendly. The partnership would give both companies a formidable presence in the business marketplace.

Apple has always favored the unconventional, as demonstrated in 2014 when Tim Cook hired a fashion executive as senior VP of retail and online stores. The position was created especially for Angela Ahrendts, the first woman to join Cook's executive team, and the first person ever assigned to oversee all 400-plus Apple

Right: Marissa Mayer, Yahoo!'s CEO and president, on the Google campus in 2008.

stores as well as online retail efforts. Cook's faith in Ahrendts was not unfounded. In her eight-year tenure as CEO of Burberry, Ahrednts rescued the struggling fashion house and nearly tripled its revenues. Not only that, but she revitalized the culture of the brand, giving it a modern cool-factor, and repositioned it as an innovator in the fashion industry. What Ahrendts had, Apple wanted. She and her work history embodied a combination of accessibility and aspiration. At her new position, Ahrendts became responsible for expanding Apple stores and redefining the customer experience.

Home Shopping Network (HSN) created the position of chief information officer for Karen Etzkorn in 2013. The move was a sign of progress for the 30-year-old multichannel retailer. Etzkorn was brought in to help develop and implement new technologies for the retailer. Since seeing that online and mobile sales made up 12 percent of HSN's overall business in 2014, she has worked toward creating an interactive experience across all channels—TV and online via tablets, mobile phones, and computers. Calling this "boundaryless retail," she hopes it will usher HSN into a new era of technology-driven commerce.

In 1997 Lori Lee joined AT&T as chief marketing officer of the Small Business division. She later moved into the company's most profitable section, Home Solutions, where she was named senior executive VP in charge of strategy, marketing, and operations for IP broadband, home phone, customer information services, and U-verse broadband and video services. Now she is tasked with leading Project Velocity IP, AT&T's multibillion-dollar plan to build competitive technology to surpass cable.

Executive women are gaining more high-powered opportunities than ever before, but it's an ongoing battle that requires the support of a strong community of women. To that end, Canadian entrepreneur Shaherose Charania founded Women 2.0, a global network and media company that creates content and organizes events with the goal of helping women in tech. Bridget van Kralingen formed Catalyst, a nonprofit organization that works for women in business. With more women supporting each other in a male-dominated industry, and more men also joining the fight to end gender inequality, the future of women in tech looks brighter.

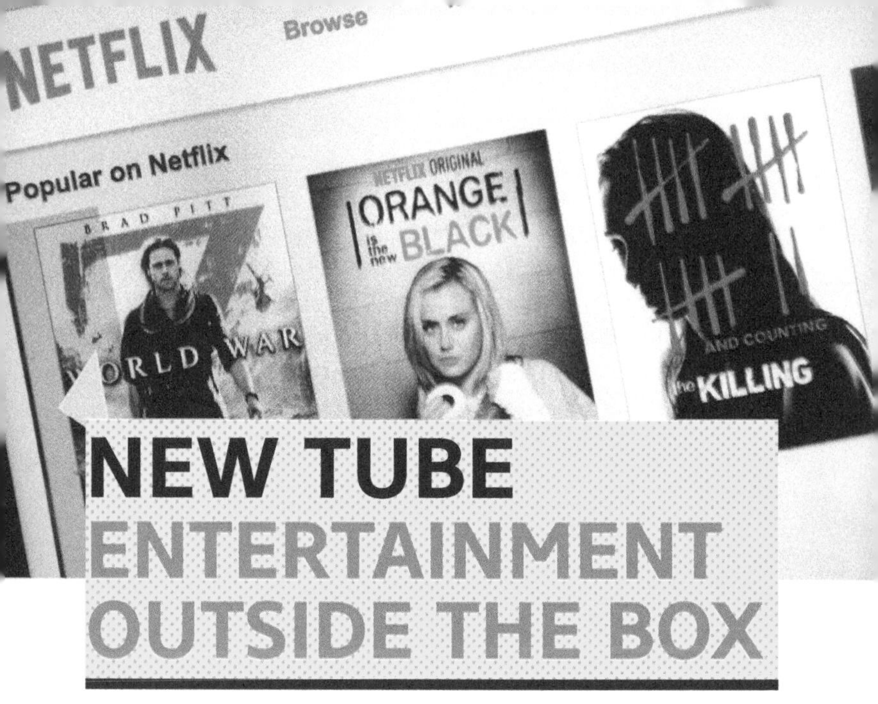

NEW TUBE ENTERTAINMENT OUTSIDE THE BOX

TECHNOLOGY CHANGES HOW, WHAT, AND WHEN WE WATCH

For decades, Americans have relied on network and cable television for their home entertainment. Advertisers could count on television networks to create programming that attracted viewers, and all viewers had to do was tune in and enjoy. Since the arrival of the Internet, though, things have changed.

One of the principal culprits for this transformation was Netflix. In 1999 the company set out to make movie rentals more convenient and economical for consumers by creating a monthly subscription-based mail-delivery rental service. In 2000 it approached the Blockbuster video rental chain and offered up its nascent company for $50 million. Blockbuster declined. By 2014, Blockbuster, once the paragon video rental success, had gone bankrupt and was shuttering its last remaining stores.

Netflix didn't stop at video rentals. In 2007 it introduced its video streaming service, which was a hit with its Internet-obsessed audience. The service created a new way of enjoying videos, giving users on-demand access to movies and television shows, right on their computers. Instead of a pricey cable bundle package, Netflix offered a different—some say better—option, at just $7.99 per month. Not only was there a vast selection of movies and TV shows, users had the flexibility to watch these offerings at any time or place, so long as an Internet connection was available. By the end of 2014, the site had more than 50 million subscribers.

Netflix upped the ante when it started to produce its own exclusive, original series that they called Netflix Originals in 2013. The first was a political drama called *House of Cards*, starring Academy Award–winning actor Kevin Spacey. It was a breakout hit. A darkly funny comedy-drama set in a women's

AMERICAN HABITS

Netflix is in 36 percent of all American households, while Amazon is in 13 percent, and Hulu Plus is in 6.5 percent. The average American adult watched nearly five hours of television each day in 2014, according to Nielsen ratings.

prison, *Orange is the New Black*, captivated audiences later that year, too. Acclaimed by critics and viewers alike, Netflix Originals had another benefit, too: there were no advertisers to please—only the subscribers. This gave the shows' producers the freedom to give viewers just what they wanted—compelling stories with high production value. Another difference was the way the shows were released; instead of releasing one episode per week, as was the standard practice in network and cable television, Netflix released an entire season at once. This was true for all of the television shows offered on Netflix, not just its own original series. As a result, users could watch episodes back to back. The phenomenon of "binge-watching" was born this way. It was instant gratification.

> *YouTube played a huge role in democratizing media—not just in terms of user-generated content, but with news content and live broadcasts as well.*

Hulu is another platform revolutionizing home entertainment. Launched in 2008, Hulu grants access to full episodes of shows from networks such as ABC, Fox, NBC, and The CW the day after they aired. The service is free for those willing to watch sit through commercials during the stream. In 2010 it also launched a subscription service, Hulu Plus, expanding its library of content to full seasons of shows as well as feature films, with minimal advertisements. Within one year, 1.5 million people had subscribed, and by 2014, 6 million. Hulu also gives users some control over their advertising experience. It uses an algorithm system that connects advertisers with compatible audiences. In-stream purchase systems will soon be padded, too, giving viewers a chance to shop directly from the streaming window. Like Netflix, Hulu produces exclusive, original series as well.

Amazon has a similar model in place with Amazon Instant Video and it offers better value for its users since the service is free for Amazon Prime customers and discounted for groups like students. Because Amazon has its own collection of digital devices like the Fire TV and Fire TV Stick, it has optimized its streaming video services to include extra features like voice activated search commands.

The advent of on-demand subscription video services has big networks brainstorming ways of diversifying access to their programming. In 2015 the premium cable network HBO announced it would offer a streaming package called HBO Now for Apple devices and that it wouldn't require a cable subscription.

For the millennial generation less time is spent in front of the television, and more time is spent on the computer not only because of services like Netflix and Hulu, but also because of websites like YouTube. When one of its founders, Jawed Karim, posted a video of himself at the San Diego Zoo in 2005, early adopters quickly followed suit, uploading more than 65,000 videos and generating more than 100 million views per day, for an average 20 million unique visitors per month. A year later, YouTube entered into a marketing and advertising partnership with NBC, before being acquired by Google a few months after that. At the end of the year, *Time* magazine named user-created media the "Person of the Year," featuring YouTube as the emblem of this fresh form of information and entertainment.

It wasn't long before YouTube had morphed into a media platform with relevance. In order to attract a faithful audience, users started to take their productions more seriously, giving their content a voice and a point of view, and making efforts to improve the visual quality. YouTube has played a huge role in democratizing media—not just in terms of user-generated content, but with news content and live broadcasts as well. The young adult demographic uses YouTube to consume news media, presidential debates, and stories about world events—although they watch plenty of YouTube's more mind-numbing content as well. Its greatest contribution, though, is that YouTube offers a place where the user can explore and learn about anything, from how-to videos to music performances and confessional video blog posts.

RISING STAR

YOUTUBE LAUNCHES NEW CAREERS FOR MILLIONS

What began as a start-up operating above a pizzeria and Japanese restaurant has now become one of the world's largest platforms for achieving stardom. Before YouTube, fame belonged solely to the rich and famous. Since the arrival of YouTube, viewers have fallen in love with the knowledge that the person on screen, in many cases, is just like them. Some YouTubers work hard to create videos that attract a loyal audience, while others become famous overnight because, for one reason or another, their video goes viral and catapults them in the news.

Almost anything is fair game when it comes to YouTube fame—makeup tutorials, product reviews, cat ladies, even the infamous "Gangnam Style" music video by the Korean singer Psy. Due to the abundance of traffic, advertisers have descended on YouTube, too, making it possible for popular YouTube personalities to make a living off of their videos. To make sure everyone cashes in, YouTube CEO Susan Wojcicki is exploring new ad formats, campaigns, and market share for the company.

WAYS WE MOVE

RAILROADS TO RIDESHARE

NEW BEGINNINGS IN TRANSPORTATION

For decades, bridges, tunnels, and thousands of miles of track were laid across the continent. By 1890, more than half a billion passengers traveled on the Pacific Railroad, along with nearly 700 million tons of cargo. Commerce was no longer confined to areas near major waterways and or coastlines. Towns and cities started to develop in the Midwest. Agriculture began thriving in new lands, and crops were easily transported to other parts of the country, even to ships that could transport goods around the world. During the 1920s, railroads started experimenting with new forms of power. Steam engines required lots of maintenance and labor, and also emitted large amounts of pollutants. New diesel locomotives were more thermally efficient and reduced maintenance and labor costs.

After the invention of the Model T, Henry Ford desired a quicker, more economical way to manufacture automobiles for the masses. In 1913, he developed a moving-chassis assembly line system inspired by the continuous-flow production approach that was used at flour mills, breweries, canneries, and bakeries. Ford's assembly line system used rope-and-pulley-powered conveyor belts on which workers built car parts like motors and transmissions. This technology allowed workers to complete an automobile in just three hours. In 1924, Ford celebrated the completion of the 10-millionth Model T. By 1930, more than half of all families in the United States owned a car. Motorized buses replaced electric trolleys, the original public transportation within cities. In 1956 President Dwight D. Eisenhower commissioned an Interstate Highway system. The major work was completed by 1990.

Fast-forward to 2015—hybrid and electric cars are slowly phasing out gas-powered vehicles. Energy efficiency is more of a concern than ever. Tesla Motors says it will soon unveil its "self-driving" car with driver-assist and auto-steering capabilities. In recent tests, self-driving cars have driven better and more safely than both stunt and regular drivers. Sensors give the car a 360-degree "view" of its surroundings, with quicker reaction times than human reflexes.

Today, drivers enjoy more freedom and control than ever on the open road. Although GPS systems were introduced in the United States around 1990, people still relied on paper maps and printed MapQuest directions into the new millennium. Around 2001, personal GPS systems started to become more affordable, and traveling by car became exponentially more flexible with the ability to transmit real-time directions. By 2008, smartphone map applications such as Google Maps and Waze, made on-the-go GPS service even more accessible and reliable. For many, it's hard to imagine driving without GPS.

Ridesharing is the newest technology shaking up transportation. Uber set out to recreate the elegant car service experience for the people of San Francisco in 2010. With its smartphone app, users can enter in their credit card information and summon a car with the tap of a button, often getting picked up within minutes. The payment is automatically processed at the end of the ride so there's no need for cash transactions. By 2014, the service had become available in 140 cities in 40 countries throughout the world. In New York, fares have become cheaper than that of a regular yellow taxi. Uber has been hailed by users for its convenience and largely criticized by taxi companies and city officials. The San Francisco Municipal Transportation Agency reported in 2014 that taxi services had seen a 65 percent drop in business since January 2012.

Lyft has been a bit less controversial, for now only serving select cities within the United States. Where Uber has created

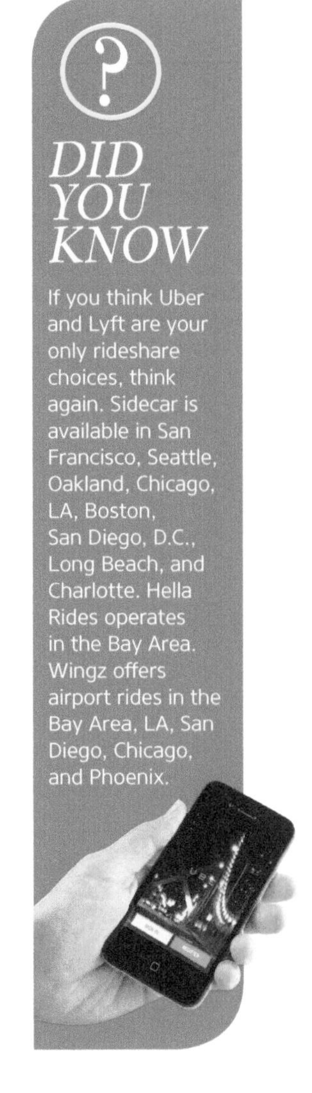

DID YOU KNOW

If you think Uber and Lyft are your only rideshare choices, think again. Sidecar is available in San Francisco, Seattle, Oakland, Chicago, LA, Boston, San Diego, D.C., Long Beach, and Charlotte. Hella Rides operates in the Bay Area. Wingz offers airport rides in the Bay Area, LA, San Diego, Chicago, and Phoenix.

a high-roller image, Lyft has opted for a more personal approach. Though it offers the same "car service" model as Uber, minus the black car opulence, it also offers Lyft Line, which groups riders who are going on similar routes into carpools. In an attempt to make ridesharing sociable rather than transactional, the company encourages conversation among drivers and passengers, going as far as launching optional user profiles within the app, giving people a chance to get to know each other prior to a ride. It's the antithesis of the self-driving car, and it remains to be seen what Americans will prefer in the long run.

Ridesharing is just another example of how technology is adapted to address human interests and social change. Services like Uber and Lyft threaten to put taxicabs out of business the same way busses phased out trolleys decades before, and diesel engines replaced steam engines before that. But if fully autonomous cars become a reality, it's possible that ride-sharing drivers, too, may be out of a job.

PASSENGER BEWARE

Drivers for UberX (Uber's budget-friendly option) are not licensed chauffeurs. They use their own cars and personal insurance policies and are not required to obtain commercial liability insurance. Although Uber claims it performs the most stringent background-screening process, many government officials are advocating for ride-share companies to adopt the same rigorous background checks as taxi drivers.

THE FUTURE IS NOW SUPERHUMAN ADVANCES IN MEDICINE

CHANGING THE FACE OF HEALTH CARE

American medicine has a reputation for focusing more on treatments that cure diseases rather than measures that prevent them. Tech companies, however, are focusing on prevention. Successful technological advancements in improving human health will, no doubt, reshape the entire health system as we know it.

Whole genome sequencing may be the key to identifying risks of developing diseases early on and require just a small amount of blood or saliva. Genetic engineering is being researched as a possibility for creating healthier humans by altering the DNA in an egg or embryo. It's controversial, however, given that the research has not progressed enough to identify all of the potential

Above: A neuron and gila as seen under microscope at 60x magnification.

consequences, such as the accidental creation of new diseases. There are also ethical concerns since this technology could theoretically allow people to create superhumans. Considering the massive expense of the technology, only a small percentage of people could afford this "advantage." Jennifer Doudna, a professor at the University of California, Berkeley, has discovered a genetic tool she calls CRISPR—Clustered Regularly Interspaced Short Palindromic Repeats—which could make the process of altering genes much less expensive. She discovered that bacteria have special enzymes that can cut open the DNA of an invading virus and alter the DNA at the cut, killing the virus. She deduced that those enzymes could work on any DNA, making it feasible to treat genetic blood diseases by tailoring therapies using the patient's own cells.

Ultrasound is still the most frequently used type of imaging test. For a long time, it was impossible to conduct these tests outside of a hospital because the mainframe machines were so large. Vscan has created a palm-sized unit that shows high-quality images of internal organs, displaying in real-time, the movement and color-coded images of blood flow against a backdrop of black-and-white anatomical images. The unit brings non-invasive point-of-care diagnostics to the bedside, and ambulance, as well as to remote or underserved areas in the United States and other parts of the world.

Brain/computer interfaces (BCI) are communication links between the human brain and an external device. Since the brain produces electrical signals, those signals may be used to control a mechanical device. A system is being developed to help rehabilitate people who have lost motor skills, such as stroke patients, and people who need assistance in controlling prosthetics. Invasive systems involve surgery to implant electrodes on the surface of the brain, while noninvasive systems use electrodes in the form of a gel placed on the scalp. The electrodes pick up brain signals and carry them through hardware to a computer, which translates them into commands. Currently, a consortium at Brown University called BrainGate is developing a wireless system. The plan is to create a small device that can be attached to the skull and wired to electrodes in the brain.

Robotics are becoming more prevalent in medical research and procedures. The da Vinci Surgical System uses magnified 3D high-definition vision and tiny instruments that possess greater dexterity and control than human wrists and hands. Although the machine's movements are still controlled by a real surgeon, they are executed more precisely by the machine. The Veebot system offers a way to automate blood draws. A patient puts their arm through a loop, and an inflatable cuff tightens around the arm. A camera uses ultrasound to confirm sufficient blood flow, then the robot aligns the needle and sticks it in. In prosthetics, scientists have been able to plug electrodes into

A WORD ABOUT

GE Healthcare partnered with the American Society of Echocardiography to bring free heart echocardiograms to a rural community in northwest India. More than 1,000 people who were suspected of having heart problems were diagnosed using the Vscan. The images were uploaded to the Internet, giving the patients access to health professionals around the world who could help.

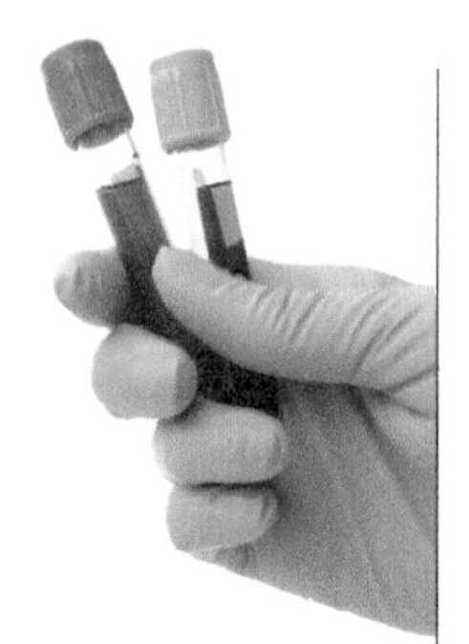

THERANOS

Elizabeth Holmes founded Theranos health technology and medical laboratory in Palo Alto, California in 2003 after dropping out of Stanford the same year. Holmes developed a streamlined way to test a single drop of blood with a simple, painless prick. The Theranos blood testing platform uses microfluidics technology, saving on commercial costs as well as using less time than traditional tests.

nerves, allowing an amputee to control the strength of a prosthetic's grip and to distinguish the shapes and stiffness of objects. Electromyography muscle sensors do a similar thing. A device designed by Meka Robotics uses these sensors to give an amputee drummer more stick control in his prosthetic hand. Meka has even created a robotic synchronization technology that listens to music and improvises an accompanying drumbeat, playing faster and more in time than the human drummer.

New developments in 3D printing are making it possible to treat more people more effectively. With 3D design, medical imaging, and 3D printing for medical applications, surgeons are able to use patient-specific data to create surgical models from which they can capture the scope of a surgery, manipulate physical solutions, and apply them during the procedure. As a result, operating times are reduced, potential errors and complications are minimized, and overall results are better for the patient. Over the last couple of years, customizable, functional

* **Marvel gave kids** 3D-printed prosthetics modeled after their favorite superheroes.

prosthetics have become more attainable because of the ability to print with inexpensive 3D printers.

Artificial intelligence is making giant leaps in healthcare. Computers have been programmed with the ability to mine data from electronic medical records in real-time, identifying patterns that humans can't. IBM's Watson computer has been serving as a reference tool for doctors to compare their own knowledge with the most up-to-date clinical research. The computer's extensive built-in knowledge of more than 600,000 pieces of medical evidence and more than 2 million pages of medical journals, and its access to 1.5 million patient records makes it "smarter" than any human doctor.

Other technology giants are focusing artificial intelligence efforts on health and medicine as well. Apple has developed five apps that deal with the health concerns of diabetes, asthma, Parkinson's disease, cardiovascular disease, and breast cancer. Google is developing contact lenses that will measure glucose levels in tears, as well as technology that fuses nanoparticles with antibodies or proteins to detect cancers; the data collected through these technologies can then be transmitted to a computer wristwatch to notify the wearer. For now, these technologies are still in very early stages of development, but it won't be long before such advanced technology becomes the mundane.

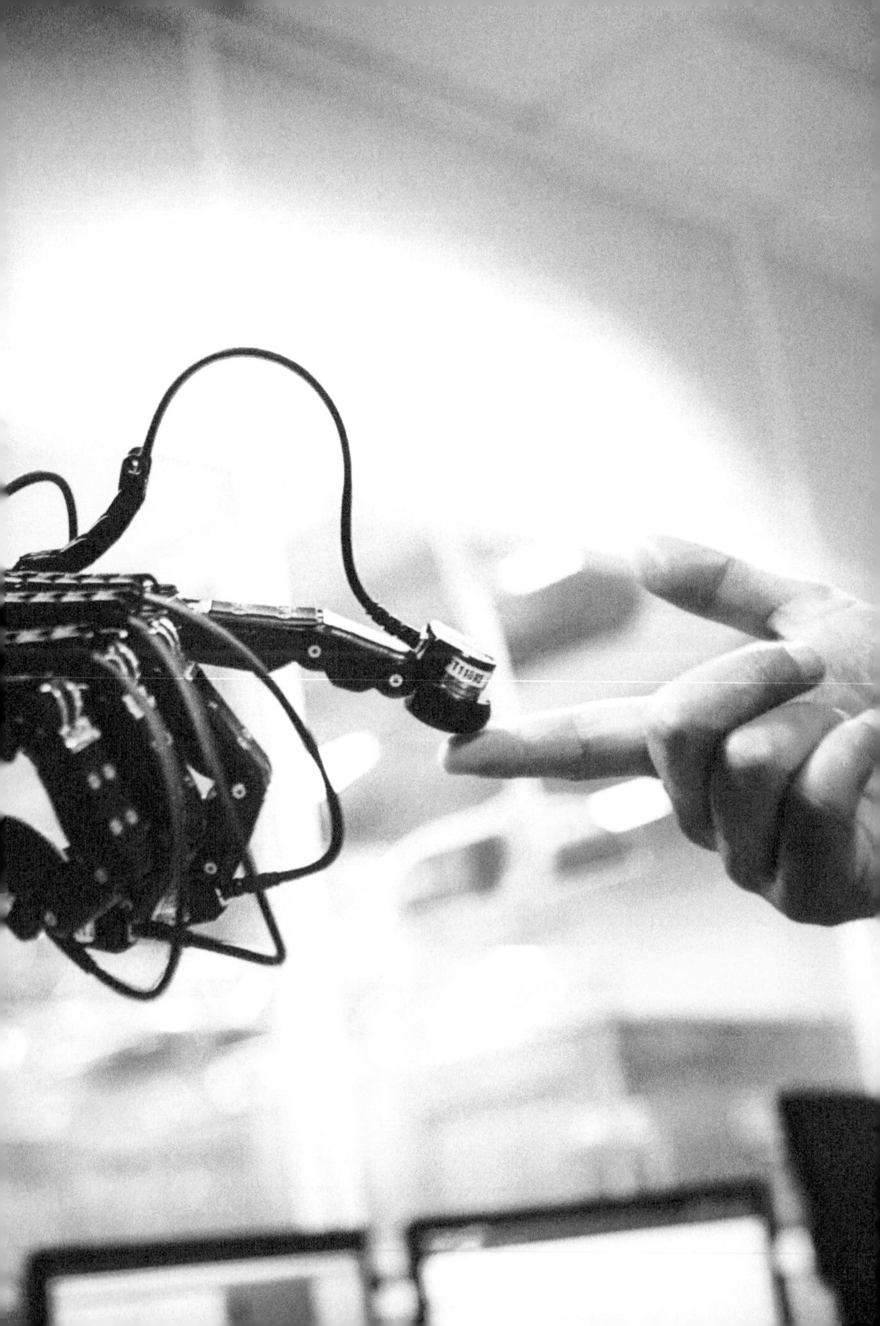

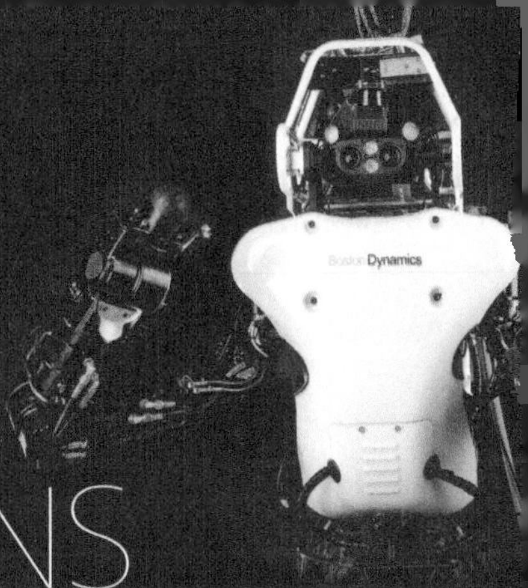

ATLAS A TITAN OF TITANS

The cover of this book features Atlas, a humanoid robot developed by Defense Advanced Research Projects Agency (DARPA) at the Pentagon. Atlas was conceived to be a hero of mankind. It will be able to perform rescue missions in environments hostile to humans. Powered by hydraulics, the highly mobile robot walks on two legs and can navigate through debris and rough terrain. With two hands, it is capable of using human tools. Its head contains a stereo vision system and a laser range finder.

In a disaster scenario, Atlas will be able to drive a vehicle, enter any premises, and complete missions involving specific manual tasks. Robots have long been a figment of the imagination, but in recent years, they've become a reality. Most robots today are not humanoid; they are usually machines used for commercial purposes. But in the future, humanoid robots will be more common. Atlas is a new breed—a robo sapien—a titan of titans.

A PARTING THOUGHT

"In string theory, all particles are vibrations on a tiny rubber band; physics is the harmonies on the string; chemistry is the melodies we play on vibrating strings; the universe is a symphony of strings, and the 'Mind of God' is cosmic music resonating in 11-dimensional hyperspace."

—MICHIO KAKU, THEORETICAL PHYSICIST

BIBLIOGRAPHY

Ackerman, Evan. "Cyborg Drumming Arm Makes Amputee into Superhuman Musician." *IEEE Spectrum*. March 6, 2014. Accessed March 26, 2015. www.spectrum.ieee.org/automaton/robotics/robotics-hardware/cyborg-drumming-arm-makes-amputee-into-superhuman-musician.

American Institute of Physics. "Imagining Ray Devices." Accessed March 19, 2015. www.aip.org/history/exhibits/laser/sections/raydevices.html.

Apple. "Apple Watch Available in Nine Countries on April 24." Accessed March 12, 2015. www.apple.com/pr/library/2015/03/09Apple-Watch-Available-in-Nine-Countries-on-April-24.html.

Arlidge, John. "Jonathan Ive Designs Tomorrow." *Time*. March 17, 2014. Accessed March 22, 2015. www.time.com/jonathan-ive-apple-interview.

AT&T. "Lori Lee." Accessed March 24, 2015. www.att.com/gen/investor-relations?pid=23854.

Auletta, Ken. "A Woman's Place." *The New Yorker*. July 11, 2011. Accessed March 24, 2015. www.newyorker.com/magazine/2011/07/11/a-womans-place-ken-auletta.

Austen, Ben. "The Story of Steve Jobs: An Inspiration or a Cautionary Tale?" *Wired*. July 23, 2013. Accessed March 22, 2015. www.wired.com/2012/07/ff_stevejobs.

Barron, James. "Before Anyone Complained About the Air-Conditioning, an Idea." *City Room* (blog). *New York Times*. July 16, 2012. www.cityroom.blogs.nytimes.com/2012/07/16/before-anyone-complained-about-the-air-conditioning-an-idea.

Bell, Gordon. "Rise and Fall of Minicomputers." Engineering Technology and History. Accessed March 18, 2015. www.ethw.org/Rise_and_Fall_of_Minicomputers.

Bell Labs. "History of Bell Labs." Accessed March 13, 2015. www.bell-labs.com/about/history-bell-labs.

Belrose, John S. "Fessenden and the Early History of Radio Science." *The Radioscientist* 5, no. 3 (September 1994).

Berners-Lee, T. "Long Live the Web." *Scientific American*. December 2010.

Bilton, Nick. "All Is Fair in Love and Twitter." *New York Times*. October 9, 2013. Accessed March 21, 2015. www.nytimes.com/2013/10/13/magazine/all-is-fair-in-love-and-twitter.html.

———. "Why Google Glass Broke." *New York Times*. February 4, 2015. Accessed March 23, 2015. www.nytimes.com/2015/02/05/style/why-google-glass-broke.html.

Bio. “Garrett Morgan.” Accessed March 12, 2015. www.biography.com/people /garrett-morgan-9414691.

Bio. “Eli Whitney.” Accessed March 16, 2015. www.biography.com/people /eli-whitney-9530201.

Bio. “Valerie Thomas.” Accessed March 12, 2015. www.biography.com/people /valerie-thomas-21341423.

BuzzFeed. “Community Post: The Complete Evolution Of Cell Phones From 1956 To The First IPhone.” January 20, 2013. Accessed April 4, 2015. http://www.buzzfeed.com/lollaparooza/the-evolution-of-cell-phones-from -1956-to-the-firs-8s5g.

Carlson, Nicholas. “The Untold Story Of Larry Page’s Incredible Comeback.” *Business Insider*. April 24, 2014. Accessed March 23, 2015. www.businessinsider.com/larry-page-the-untold-story-2014-4.

Carlson, Peter. “The Bell Telephone: Patent Nonsense?” *The Washington Post*. February 20, 2008. Accessed March 16, 2015. www.washingtonpost.com /wp-dyn/content/article/2008/02/19/AR2008021902596.html.

Carrier. “Willis Carrier.” Accessed March 15, 2015. www.williscarrier.com.

Carroll, Rory. “Bell Did Not Invent Telephone, US Rules.” *The Guardian*. June 17, 2002. Accessed March 16, 2015. www.theguardian.com /world/2002/jun/17/humanities.internationaleducationnews.

Celebrating Texas. “Bette Nesmith Graham.” Accessed March 14, 2015. www.celebratingtexas.com/tr/lsl/94.pdf.

Chu, Jeff. “Can Apple’s Angela Ahrendts Spark a Retail Revolution?” *Fast Company*. January 6, 2014. Accessed March 24, 2015. www.fastcompany.com /3023591/angela-ahrendts-a-new-season-at-apple.

CNN. “4 in 10 TV Households Also Subscribe to Netflix, Amazon, or Hulu.” Accessed March 25, 2015. www.money.cnn.com/2015/03/11/media /nielsen-report-netflix-amazon-hulu.

Cognitive Code. “SILVIA for Android.” Accessed March 13, 2014. www.silvia4u.info/silvia-for-android.

Cowan, Ruth Schwartz. *A Social History of American Technology*. New York: Oxford University Press, 1997.

Dunham, Scott. “Surgeon’s Helper: 3D Printing Is Revolutionizing Health Care.” *Livescience*. February 23, 2015. Accessed March 26, 2015. www.livescience.com/49913-3d-printing-revolutionizing-health-care.html.

Dupont, Veronique. “Netflix Has Revolutionized the TV Industry Several Times In Just 17 Years.” *Business Insider*. September 12, 2014. Accessed March 25,

2015. www.businessinsider.com/afp-netflix-the-revolution-that-changed -the-us-tv-landscape-2014-9.

Eli Whitney Museum and Workshop. "Eli Whitney: The Inventor." Accessed March 16, 2015. www.eliwhitney.org/7/museum/about-eli-whitney/inventor.

Federal Trade Commission. "In the Matter of Facebook, Inc. a Corporation." Accessed March 23, 2015. www.ftc.gov/sites/default/files/documents/cases/ 2012/08/120810facebookcmpt.pdf.

Forbes. "The World's 100 Most Powerful Women." Accessed March 25, 2015. www.forbes.com/profile/susan-wojcicki.

Forbes. "The World's Most Powerful People." Accessed March 21, 2015. www.forbes.com/profile/ginni-rometty.

Forbes.com. "The Most Powerful Women in Tech". Accessed May 11, 2015. http://www.forbes.com/sites/zheyanni/2014/05/28/the-most-powerful -women-in-tech-2014/

Foster, Tim. "Michael Dell: How I Became an Entrepreneur Again." *Inc.* Accessed March 21, 2015. www.inc.com/magazine/201407/tom-foster/michael-dell -on-transformating-dell.html

Friend, Tad. "Hollywood and Vine." *The New Yorker.* December 15, 2014. Accessed March 25, 2015. www.newyorker.com/magazine/2014/12/15/hollywood-vine.

Geoghegan, Tom. "Twitter, Telegram, and Email: Famous First Lines." *BBC News.* March 21, 2011. Accessed March 14, 2015. www.bbc.co.uk/news /magazine-12784072.

Globe Newswire. "HSN, Inc. Names Karen Etzkorn Chief Information Officer." Accessed March 24, 2015. www.globenewswire.com/news-release /2012/12/04/509335/10014426/en/HSN-Inc-Names-Karen-Etzkorn-Chief-Information-Officer.html.

Gomes, Lee. "Sun Microsystems' Rise and Fall." *Forbes.* March 20, 2009. Accessed March 19, 2015. www.forbes.com/2009/03/18/sun-microsystems -internet-technology-enterprise-tech-sun-microsystems.html.

Google. "Our History in Depth." Accessed March 21, 2015. www.google.com/about/company/history.

Grant, John. "Experiments and Results in Wireless Telephony." *The American Telephone Journal* (January 26, 1907): 49–51.

Gustin, Sam. "Apple and Google Call a Truce in Patent Wars." *Time.* May 16, 2014. Accessed April 10, 2015. www.time.com/103640/apple-google-patent-truce/.

Handy, Galen. "Electric Cars." Edison Tech Center. Accessed March 13, 2015. www.edisontechcenter.org/ElectricCars.html.

Harris, Scott Duke. "The Best Thing Since Sliced Bread?" *Mercury News*. July 2, 2007. Accessed March 12, 2015. www.mercurynews.com/business/ci_ 6280200.
Harvard University. "Early Wired Telegraph." Accessed March 15, 2015. www.people.seas.harvard.edu/~jones/cscie129/images/history/von_Soem.html.
Hernandez, Daniel. "Artificial Intelligence Is Now Telling Doctors How to Treat You." *Wired*. June 2, 2014. Accessed March 26, 2015. www.wired.com/2014/06/ai-healthcare.
History. "Ford's Assembly Line Starts Rolling." Accessed March 25, 2015. www.history.com/this-day-in-history/fords-assembly-line-starts-rolling.
History.com. "Thomas Edison." Accessed May 11, 2015. http://www.history.com/topics/inventions/thomas-edison
IBM. "The Apollo Missions." Accessed March 20, 2015. www-03.ibm.com/ibm/history/ibm100/us/en/icons/Apollo.
IBM. "IBM." Accessed March 20, 2015. www-03.ibm.com/ibm/history/ibm100/us/en.
Id Software. "Doom Press Release." January 1, 1993.
Intel. "The History of Intel, 30 Years of Innovation." Accessed March 19, 2015. www.landley.net/history/mirror/intel/cn71898a.htm.
Isaacson, Walter. Steve Jobs. New York, NY: Simon & Schuster, 2011.
Kata, Lauren. "Marion O'Brien Donovan Papers." National Museum of American History. 2000. Accessed March 14, 2015. www.amhistory.si.edu/archives/d8721.htm.
Kessler, Glenn. "A Cautionary Tale for Politicians: Al Gore and the 'Invention' of the Internet." *The Washington Post*. November 4, 2013. Accessed March 13, 2014. www.washingtonpost.com/blogs/fact-checker/wp/2013/11/04/a-cautionary-tale-for-politicians-al-gore-and-the-invention-of-the-internet.
Killingly Historical Society. "Mary Dixon Kies, America's First Female Patent Holder." Accessed March 24, 2015. www.killinglyhistory.org/online-journals/online-journal-vol-7-2005/23-mary-dixon-kies-americas-first-female-patent-holder-.html.
King, Leo. "Apple-IBM Partnership: Enough to Solve Enterprise iOS Fears?" *Forbes*. July 16, 2015. Accessed March 24, 2015. www.forbes.com/sites/leoking/2014/07/16/apple-ibm-partnership-enough-to-solve-enterprise-ios-fears.
Lashinsky, Adam. "The Future Will Be Quantified." *Fortune*. June 2, 2014. Accessed March 24, 2015. www.fortune.com/2014/06/02/connected-bridget-van-kralingen-ibm.
Lean In. "Ursula M. Burns." Accessed March 24, 2015. www.leanin.org/stories/ursula-burns.

Lehman, Milton. *This High Man: The Life of Robert H. Goddard*. New York: Farrar, Strauss, and Co., 1963.

Lev-Ram, Michal. "The powerful woman behind Intel's new $300 million diversity initiative." *Fortune*. January 12, 2015. Accessed March 21, 2015. www.fortune.com/2015/01/12/intel-diversity/.

McCullough, David. "Samuel Morse's Reversal of Fortune." *Smithsonian Magazine* September 2011. Accessed March 15, 2015. www.smithsonianmag.com /history/samuel-morses-reversal-of-fortune-49650609/?no-ist=&onsite _medium=internallink&page=1.

McGrath, Charles. "Everything and the Kitchen Sink: The Memoir of a Dishwasher." *The New York Times*, May 23, 2007. www.nytimes.com/2007/05 /23/books/23-suds.html?pagewanted=all.

Miller, Michael J. "Creating the 8080: The Processor That Started the PC Revolution." Forward Thinking, PC Mag. December 18, 2014. Accessed March 18, 2015. www.forwardthinking.pcmag.com/none/330501-creating-the-8080 -the-processor-that-started-the-pc-revolution.

Morse, Samuel F. B. *Samuel F. B. Morse: His Letters and Journals*. FQ Books, 2010.

NASA. "TIROS." Accessed March 14, 2015. www.science1.nasa.gov/missions/tiros.

National Institutes of Health. "New NIH-funded Ultrasound Technology is Changing Lives around the World." Accessed March 26, 2015. www.nlm.nih. gov/medlineplus/magazine/issues/winter13/articles/winter13pg26-27.html.

National Park Service. "A Few Gifted Men Who Worked For Edison." Accessed March 12, 2015. www.nps.gov/edis/learn/kidsyouth/the-gifted-men-who -worked-for-edison.htm.

National Park Service. "Volta Laboratory & Bureau." Accessed March 13, 2015. www.nps.gov/nr/travel/wash/dc14.htm.

Nikolewski, Rob. "The Uber Effect: Why cab companies hate ridesharing." *California Watchdog.org*. September 29, 2014. Accessed March 26, 2015. watchdog.org/173904/uber-effect-ridesharing.

Nobelprize.org. "The History of the Integrated Circuit." Accessed March 18, 2015. www.nobelprize.org/educational/physics/integrated_circuit/history.

Noguchi, Yuki. "Duke Energy CEO: 'I Don't Think of Myself as a Powerful Woman.'" NPR. October 21, 2014. Accessed March 24, 2015. www.npr. org/2014/10/21/357818516/duke-energy-ceo-i-dont-think-of-myself -as-a-powerful-woman.

NPR. "'Lean In': Facebook's Sheryl Sandberg Explains What's Holding Women Back." Accessed March 24, 2015. www.npr.org/2013/03/11/173740524/lean -in-facebooks-sheryl-sandberg-explains-whats-holding-women-back.

Olsen, Stefanie. "Sun's John Gage joins Al Gore in clean-tech investing." *CNET*. June 9, 2008. Accessed March 26, 2015. www.cnet.com/news/suns-john -gage-joins-al-gore-in-clean-tech-investing.

O'Neal, James E. "Fessenden—The Next Chapter." Radioworld. December 23, 2008. Accessed March 17, 2015. www.radioworld.com/article/fessenden -%E2%80%94-the-next-chapter/273.

———. "Fessenden: World's First Broadcaster?" Radioworld. October 25, 2006. Accessed March 17, 2015. www.radioworld.com/article /fessenden-world39s-first-broadcaster/15157.

Oracle. "Oracle's History: Innovation, Leadership, Results." Accessed March 21, 2015. www.oracle.com/us/corporate/history/index.html.

Palca, Joe. "A CRISPR Way to Fix Faulty Genes." *NPR*. June 26, 2014. Accessed March 26, 2015. www.npr.org/blogs/health/2014/06/26/325213397/a -crispr-way-to-fix-faulty-genes.

Parker, Ian. "The Shape of Things to Come." *New Yorker*. February 23, 2015. Accessed March 22, 2015. www.newyorker.com/magazine/2015/02/23 /shape-things-come.

PBS. "Transistorized!" Accessed March 17, 2015. www.pbs.org/transistor /album1/.

PC-History. "How the Altair Began." Accessed March 18, 2015. www.pc-history. org/altair.htm.

Peters, Betts, and Melanie Fried-Oken. "FYI: Brain-Computer Interface." ALS Association. September 2014. Accessed March 26, 2015. www.alsa.org/als-care/ resources/publications-videos/factsheets/brain-computer-interface.html.

Pogue, David. "Lyft CEO: We Do Ride-Sharing, but 'Not for High Rollers.'" March 25, 2015. Accessed March 26, 2015. www.yahoo.com/tech/logan-green-the -cofounder-and-ceo-of-lyft-is-a-114560256229.html.

Popken, Ben. "States Warn of Rideshare Risks for Passengers." *NBC News*. May 28, 2014. Accessed March 26, 2015. www.nbcnews.com/business /consumer/states-warn-rideshare-risks-passengers-n116736.

Regalado, Antonio. "A Brain-Computer Interface That Works Wirelessly." *MIT Technology Review*. January 14, 2015. Accessed March 26, 2015. www.technologyreview.com/news/534206/a-brain-computer-interface -that-works-wirelessly.

Ritter, Karl, and Malin Rising. "2 Americans, 1 German Win Chemistry Nobel." Accessed March 13, 2015. www.apnews.excite.com/article/20141008 /nobel-chemistry-e759dff699.html.

Samsung. "Dr. Oh Hyun Kwon." Accessed March 12, 2015. www.samsung. com /us/aboutsamsung/samsung_electronics/management/president.html.
Schatzkin, Paul. *The Boy Who Invented Television*. Silver Spring, Maryland: Teamcom Books, 2002.
Scott Products. "Sensible Innovations." Accessed March 14, 2015. www.scott-brand.com/aboutus.
Seabrook, John. "Revenue Streams." *The New Yorker*, November 24, 2014. Accessed March 25, 2015. www.newyorker.com/magazine/2014/11/24 /revenue-streams.
Shontell, Alyson. "The Impressive Career of Yahoo CEO Marissa Mayer." *Business Insider*, October 9, 2012. Accessed March 24, 2015. www.businessinsider. com/marissa-mayer-ceo-of-yahoos-career-2012-10.
Smithsonian National Air and Space Museum. "Inventing a Flying Machine." Accessed March 16, 2015. www.airandspace.si.edu/exhibitions/wright -brothers/online/fly/1899/index.cfm.
Smithsonian National Museum of History. "Transportation Technology." Accessed March 25, 2015. www.amhistory.si.edu/onthemove/themes /story_50_1.html.
SPIE. "In memoriam: James L. Fergason." Accessed March 12, 2015. www.spie. org/x32291.xml.
Stein, Rob. "Scientists Urge Temporary Moratorium on Human Genome Edits." NPR. March 20, 2015. Accessed March 26, 2015. www.npr.org/blogs/health /2015/03/20/394311141/scientists-urge-temporary-moratorium-on-human -genome-edits.
Swisher, Kara. "Man and Uber Man." *Vanity Fair*, December 2014. Accessed March 26, 2015. www.vanityfair.com/news/2014/12/uber-travis-kalanick-controversy.
Sydell, Laura. "Intel Legends Moore and Grove: Making It Last." NPR. April 6, 2012. Accessed March 19, 2015. www.npr.org/2012/04/06/150057676/intel -legends-moore-and-grove-making-it-last.
Tett, Gillian. "Interview: Facebook's Sheryl Sandberg." *Financial Times*. April 19, 2013. Accessed March 24, 2015. www.ft.com/cms/s/2/da931d58-a7c2-11e2 -9fbe-00144feabdc0.html#axzz3VLVd5Pww.
Tsividis, Yannis. "Edwin Armstrong: Pioneer of the Airwaves." *Columbia* magazine. Accessed March 12, 2015. www.columbia.edu/cu/alumni/Magazine /Spring2002/Armstrong.html.
Tsokos, K. A. *Physics for the IB Diploma Full Colour*. Cambridge University Press, 2010.

The National Museum of American History. "Supermarket Scanner." Accessed March 19, 2015. www.americanhistory.si.edu/collections/search/object/nmah_892778.

US patent 1115674. Mary Phelps Jacob. "Backless Brassiere." Issued November 3, 1914.

US Patent 1233597 A. William J. Newton. "Flush switch." Issued July 17, 1917.

US Patent 3449750 A. George H. Sweigert. "Duplex radio communication and signaling apparatus for portable telephone extension." Issued June 10, 1969.

Vascellaro, Jessica E. "Facebook Grapples with Privacy Issues." *The Wall Street Journal*. May 19, 2010. Accessed March 23, 2015. www.wsj.com/articles/SB10001424052748704912004575252723109845974.

Wadhwa, Vivek. "Apple Isn't Just Satisfied Reinventing Health Care, It's Targeting Clinical Trials As Well." *Washington Post*. March 23, 2015. Accessed March 26, 2015. www.washingtonpost.com/blogs/innovations/wp/2015/03/23/apple-isnt-just-satisfied-reinventing-health-care-its-targeting-clinical-trials-as-well%E2%80%8B.

Williams, Bernard O. "Computing with Electricity, 1935–1945," PhD Dissertation, University of Kansas, 1984. University Microfilms International, 1987, p. 310.

Winkler, Rolfe. "YouTube: 1 Billion Viewers, No Profit." *The Wall Street Journal*. February 25, 2015. Accessed March 25, 2015. www.wsj.com/articles/viewers-dont-add-up-to-profit-for-youtube-1424897967.

Wohleber, Curt. "The Vacuum Cleaner." *Invention & Technology* magazine. American Heritage Publishing. Accessed March 12, 2015. www.invention-and-tech.com/content/vacuum-cleaner-0.

Wohlsen, Marcus. "Uber's Brilliant Strategy to Make Itself Too Big to Ban." *Wired*, July 8, 2014. Accessed March 26, 2015. www.wired.com/2014/07/ubers-brilliant-strategy-to-make-itself-too-big-to-ban.

Wong, Wylie. "Oracle Chief Defends Microsoft Snooping." CNET. Accessed March 21, 2015. www.news.cnet.com/Oracle-chief-defends-Microsoft-snooping/2100-1001_3-242560.html.

Yale University. "Grace Murray Hopper." Accessed March 12, 1015. www.cs.yale.edu/homes/tap/Files/hopper-story.html.

Zimmer, Michael. "Mark Zuckerberg's theory of privacy." *The Washington Post*. February 3, 2014. Accessed March 23, 2015. www.washingtonpost.com/lifestyle/style/mark-zuckerbergs-theory-of-privacy/2014/02/03/2c1d780a-8cea-11e3-95dd-36ff657a4dae_story.html.

INDEX

D

E

F

T

U

V